TRANSATLANTIC

AIRSHIPS

AN ILLUSTRATED HISTORY

DEUTSCHE ZEPPELIN-REEDEREI

TRANSATLANTIC
AIRSHIPS
AN ILLUSTRATED HISTORY

JOHN CHRISTOPHER

THE CROWOOD PRESS

First published in 2010 by
The Crowood Press Ltd
Ramsbury, Marlborough
Wiltshire SN8 2HR

www.crowood.com

British Library Cataloguing-in-Publication Data
A catalogue record for this book is available from the British
Library.

ISBN 978 1 84797 161 6

FRONTISPIECE: The cover of a publicity brochure for the *Graf
Zeppelin*.

Designed and typeset by Focus Publishing, Sevenoaks, Kent
Printed and bound in Malaysia by Konway Printhouse Sdn Bhd

Contents

Acknowledgements

Producing a book on such a wide subject as transatlantic airships has been greatly assisted with the kind help of a number of individuals and organisations who have provided invaluable information and images. I would like to express my thanks to the following:

Herman Van Dyk, who sent me information on several of the more obscure earlier transatlantic attempts; Richard Van Trueren, editor of the *Noon Balloon*, the official newsletter of the Naval Airship Association in the USA, for information and images from the flight of the US Navy's *Snow Bird*; many members of the Airship List who have helped with some useful pointers, especially Ford Ross, who very kindly provided photographs of several of the transatlantic airships, and also Rick Zitarosa of the Navy Lakehurst Historical Society; Richard Gyselynck; Alexandra and Brian Hall of Airship Ventures, which operates the Zeppelin NT in the San Francisco area; Daniel Grossman, who assisted with additional photographs of the Zeppelins; Gordon Taylor and Hybrid Air Vehicles in the UK; and Francisco Gonzalez Redondo for information on the Spanish contribution to the airship story. Thanks also to Campbell McCutcheon and The Crowood Press for their faith in this project. My appreciation also goes to the US Library of Congress and the US Navy for their enlightened attitude to providing images within the public domain.

Naturally I must express my gratitude to my wife, Ute, for her continued support, not forgetting the many hours of proof reading, and to my children Anna and Jay for their patience. They now know more about airships than any children of their age probably should.

Introduction

'We passed over a symphony of silver to golden glory as the lights of New York City scattered themselves beneath us like grains of golden stardust, tracing patterns strange and fantastic, set with the jewelled brilliancy of ruby, emerald and topaz electric signs...'

These words by Lady Drummond Hay, the journalist with Randolph Hearst's Universal News service who became such a frequent flyer that she was given the unofficial title of 'Lady Zeppelin', come to us from another time. A distant, dream-like time when aerial leviathans roamed the skies. The golden age of transatlantic airship travel may have been shortlived, spanning a few brief years from the 1920s to the 1930s, but it was one in which air travel became an unforgettable experience in itself, albeit the reserve of the wealthy and the famous. It was a world apart from the modern traveller's experience of flying in one of today's high-density passenger jets, approximately 1,000 of which cross the Atlantic every single day.

For several centuries this formidable ocean has represented a physical and symbolic divide between the old and the new worlds. Conquered by the great sailing ships of the past, it quickly became an irresistible challenge for the fledgling aviators. Many brave men died in the attempt, and only a handful succeeded at first. Yet despite the likes of Alcock and Brown, who made the first non-stop crossing by aircraft in 1919, or Charles Lindbergh, who flew from New York to Paris in the *Spirit of St Louis* in 1927, up until the Second World War the future of long-distance air travel was widely judged to lie with the airship alone. Lady

Max Pruss with Hearst Newspaper correspondents Karl von Wiegand and Lady Grace Drummond Hay, aboard the Graf Zeppelin.

Drummond-Hay's account of travelling in the *Graf Zeppelin* was written only two years after Lindbergh's epic solo flight, and in stark contrast to his ordeal of endurance she travelled in considerable comfort.

Not surprisingly, it was the schemes for regular international airship services that filled the newspapers and magazines to stir the public's imagination with their fantastic visions of futuristic airliners, while the aeroplane's early exploits were readily dismissed. As the aviator Harry Hawker commented after his failed transatlantic attempt in May 1919: 'I need hardly say that commercial aeroplanes are scarcely likely to be called upon to make non-stop flights of upwards of thousands of miles. That sort of thing can probably be better done with an airship.'

He was right, up to a point. In the late 1920s and the early 1930s the transatlantic Zeppelins were the only long-distance air carriers, operating on a scheduled transatlantic service between Germany and either New York or Rio de Janeiro. Flying on board the LZ129 *Hindenburg*, the passengers were able to relax in the spacious public rooms, they ate excellent food freshly cooked in the airship's well-equipped galley and served on fine china, and they slept in comfortable cabins. To while away the hours during a two- or three-day crossing there was a lightweight baby grand piano made of duralumin alloy in the saloon area, as well as a small bar on the lower deck, a smoking room and even a shower room. The transatlantic Zeppelins established a standard of service that other aircraft would not achieve for decades to come. As Dr Hugo Eckener, the driving force behind the Zeppelin company between the two world wars, once famously said: 'You don't fly in an airship, you go voyaging.'

So it may be surprising to note that a total of only 14 different airships made successful transatlantic crossings. Apart from the scheduled Zeppelin services, most of these were pioneering flights, while a few were for military purposes. In addition, however, there were many hopefuls, the would-be 'Pond-hoppers' who failed in the attempt. All of these have their place in the story of transatlantic airships, and this account also takes the opportunity to tell of the countless visionary schemes that never to get off the drawing board, let alone off the ground.

The Graf Zeppelin *in the shed at Friedrichshafen. Most transatlantic flights began in the cool air of the pre-dawn.*

Instructions are given to the ground-handling party to march the Graf Zeppelin *out of its shed.*

Artist Ludwig Dettmann was aboard the Graf Zeppelin on the first flight to the USA in October 1928, and he captured this scene as they flew over the lights of New York on the return leg.

A few years ago an exhibition was held in Philadelphia to celebrate the work of the artist Arthur Radebaugh, who produced many of the futuristic illustrations for various American magazines such as *Popular Mechanics* and *Popular Science*. With his wonderful and extraordinary flying cars, high-speed monorails, atomic-powered vehicles, exotic submarines and, of course, improbable airships, Radebaugh portrayed a future brimming with technological exotica. One of the most prominent images published at the time of the exhibition was of a giant airship moored to the top of a space-age pylon which spanned a city and pierced the sky. Yes, it was a vision of a future that was never to be, and maybe a better one than we actually got, but the airship has always represented modernity and forward thinking. Jules Verne and H.G. Wells both conjured up incredible flying machines in their writings. Wells in particular anticipated the Zeppelin raids on London in the First World War when he depicted a vast fleet of German airships attacking New York in his 1907 novel *The War in the Air*.

No wonder, then, that in the surreal world of lighter-than-air flying the distinctions between fact and fantasy have often become blurred, one spilling over into the other. But just take a look at the great airships, especially the photographs of the superb Bauhaus-inspired interiors of the *Hindenburg* or its sister ship, the LZ130 *Graf Zeppelin II*, and they are just as incredible as any of Radebaugh's illustrations in their own right. This, I believe, is one of the fundamental reasons why so many people are smitten by airships in the first place. It is because they are both improbable and fantastic at the same time that they continue to captivate the imagination, especially through the all-too-brief period of the great transatlantic airships which represents the pinnacle of their achievements to date.

CHAPTER 1

Flights of Fancy

At the dawn of the twentieth century there were two distinct branches of aviation. The lighter-than-air balloonists or aeronauts, who had been at it since the Montgolfier brothers' first manned balloon had risen from the Chateau de la Muette, Paris, in 1783, and the upstart heavier-than-air aviators who were making their first hops into the air in their wonderful flying machines 120 years later. By 1900 only a handful of experimental airships had been flown, and the future direction of aviation remained wide open and the subject of much speculation.

The balloonists

In many ways it is surprising that the balloonists had not managed to conquer the Atlantic, and in fact they would turn out to be the 'also-rans' of this race, although it was not through any lack of ambition or the willingness to give it a go. The British aeronaut Charles Green is credited as the first to turn his attention to a transatlantic flight. A brave undertaking, considering that the greatest distance achieved thus far, and by Green himself, was some 380 miles (612km) on a flight from London to Nassau,

Germany, in 1836. By observing the movement of the clouds and the weather systems Green identified a reasonably consistent west-to-east transatlantic airflow at higher altitudes, roughly corresponding to the lower-level trade winds familiar to sailors. Twice he advertised for 'wealthy patrons of the art' to underwrite the project, and twice they failed to materialize. On the other side of the Atlantic the American balloonist John Wise had observed the same airflows. 'A current from west to east in the atmosphere is constantly in motion within the height of 12,000ft (4,000m) above the ocean.' He was also a firm believer in the viability of the balloon as a means of transatlantic transportation to rival the steamships, having once stated: 'Our children will travel to any part of the globe without the inconvenience of smoke, sparks and seasickness, and at the rate of 100 miles per hour [160km/h].'

With backing from a wealthy Vermont businessman, Wise took his appropriately named *Atlantic* balloon for a pre-oceanic test flight. At dusk on 1 July 1859 the 120ft (37m)-high gas balloon ascended from St Louis, and by the following afternoon it was to be found halfway up a tree near Lake

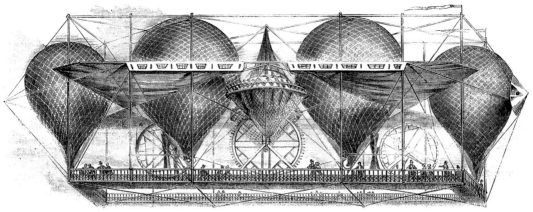

M. PETIN'S AERIAL LOCOMOTIVE MACHINE.

Published in The London Journal *in 1850, Monsieur Petin's Aerial Locomotive Machine may be far-fetched, but it does anticipate the concept of a train of balloons strung together.*

Ontario. Wise and his three companions, fellow aeronaut John La Mountain, his backer O.A. Gager and newspaper reporter William Hyde, had encountered gale-force winds of up to 90mph (145km/h) which had slammed the balloon through the forest, ripping it to tatters. They had set a new distance record, having flown an incredible 809 miles (1,300km) in just under twenty hours, more than doubling Green's record but woefully short of that required to reach Europe.

John La Mountain rebuilt the *Atlantic*, but he fared little better on the test flight. Launching from New York, he ended up stranded in the Canadian wilderness for four days. Next up was fellow-American Thaddeus Lowe, who launched from Cincinnati on 19 April 1861 and landed some 350 miles (563km) later at Unionsville, South Carolina, where he was promptly arrested as a Yankee spy. The American Civil War put paid to any further

serious attempts by the balloonists for some time, and incredibly the Atlantic would remain beyond their reach until the flight of the *Double Eagle II* in 1979, nearly 200 years after the birth of ballooning.

Dawn of the airships

Even though the balloonists would have to wait their time, their activities served to inspire a number of inventive minds in the latter half of the nineteenth century with the concept of the 'dirigible' balloon, from the French for 'steerable'. In other words an airship. It is worth noting that the terminology remained very fluid in those early days, and the term air-ship or airship continued to be applied as much to the creations of the heavier-than-air wood and fabric brigade as to the lighter-than-air fabric and hydrogen fliers for some time.

As early as 1784, the year after the Montgolfiers' first manned balloon had flown, a French army engineer named Jean Baptiste Marie Meusnier produced drawings of an inspired concept for an airship which incorporated an elongated envelope to be filled with hydrogen, with a gondola or boat-shaped car suspended beneath it. To maintain the envelope's internal pressure and to prevent wrinkling, Meusnier proposed an internal air-filled bladder or 'ballonet' (small balloon) which could be expanded or contracted without any loss of the all-important lifting gas. This was in essence is what we now know as a 'pressure' airship, because it maintained its shape entirely through the maintenance of its internal pressure. Meusnier's design never progressed beyond the concept stage, but others would take a more practical approach, and in 1852 Frenchman Henri Giffard took to the air in a steam-powered airship. While Giffard demonstrated that he could steer the cigar-shaped craft through the air, the steam engine was lacking in power and represented something of a health hazard with its proximity to the flammable hydrogen.

The latter half of the nineteenth century proved to be a fertile breeding period for countless inventors or engineers, and the files of patent offices were brimming with designs for wonderful aerial conveyances and vessels in all shapes and sizes. Many were concerned with methods of propulsion, from human-powered 'flappers', paddle wheels and cylindrical helical screws to various steam and gas engines. One 1861 proposal suggested providing propulsion from the effects of combining ammonia and carbonic acid. Many inventors drew their inspiration from the observation of fish moving through water, as in Micaiah Hill's 1880 Provisional Patent for a 'Navigable Balloon': 'A fish-shaped balloon is propelled by compressed air, which is allowed to escape by suitable orifices, situated about one-third of the length of the balloon from the head. A flexible tail is employed for steering.'

Thankfully, Hill's flatulent fish was not

the shape of things to come, and the more familiar cigar shape was becoming increasingly prevalent. In another patent from 1880, William Robert Lake foresaw an elongated balloon made rigid by an internal framework of thin metal tubing, with the interior divided into separate partitions. He was not too far off the mark.

Undaunted, others continued their own lines of development, such as Dr A. de Baussett, whose big idea was to build a steel airship to be 'thoroughly braced inside' to resist atmospheric stresses when a partial vacuum was obtained inside. Powered by electric motors, the *Ben Franklyn*, as it was to be known, would be big enough to carry 200 people or fifty tons of mail to 'swoop' through the air at a rate of at least 70mph (113km/h). As a report in *The New York Times* of 16 August 1889 commented:

> The possibilities of Dr de Bausset's air ship are hardly to be grasped by the ordinary mind. It will, according to the timetable just issued, land us in Liverpool in forty-three hours, or in Chicago in thirteen hours... The Emperor William can sail over to Washington... and be back in Berlin in less than four days. The possibilities are as boundless as space.

Dr de Bausset's concept might have been flawed, but in his endeavours he had also discovered a particular obstacle that has taxed the airship's proponents ever since; raising the cash he needed. In many ways his chosen method, through a popular national subscription, was a remarkable prophesy of future events.

Still the quest for a suitable power source continued, and it was not until the advent of the internal combustion engine that the airship builders finally had a propulsion system that was both powerful, reliable and lightweight. In France, Brazilian-born aeronaut Alberto Santos-Dumont made a name for himself by developing a series of small pressure airships using petrol engines. In 1901 he triumphantly claimed the Deutsch Prize by making the first flight from St Cloud, on the outskirts of Paris, to encircle the Eiffel Tower and return in under ten minutes. In Germany another pioneer was taking a different approach to the design of a dirigible, although it is interesting to note that Zeppelin was not working in a vacuum as the air was, figuratively at least, awash with ideas for navigable balloons, as we have seen from this small selection of nineteenth-century patents.

Zeppelin's big idea

Count Ferdinand von Zeppelin, a retired Prussian cavalry officer, had encountered observation balloons during the American Civil War. His concept was to create an airship as big as an ocean liner by taking a string of gas balloons and containing them within a lightweight metal skeleton, or framework, comprised of circular rings held in position by longitudinal girders within a protective fabric outer cover. This was the 'rigid' airship. In contrast to the pressure ships, the main advantage was that the gas cells could contract or expand as required without compromising the airship's overall shape. Having retired from the Prussian Army under a cloud, Zeppelin was determined to restore his reputation and to build a practicable airship to secure a military advantage for his country.

On 2 July 1900 Zeppelin's prototype airship, the Luftschiff Zeppelin 1, or LZ1, took to the air from a floating hangar near Friedrichshafen, on the shores of Lake Constance. It was designed by Dr Theodor Kober, assisted by the brilliant young engi-

Ferdinand Adolf Heinrich August Graf von Zeppelin, 1838–1917, founder of the Zeppelin company and widely regarded as the father of the rigid airship.

neer Ludwig Durr, who went on to design all of the Zeppelin airships. The LZ1 was 420ft (676m) long and had a capacity of 399,000cu ft (11,290cu m). That first flight lasted only 18 minutes, and in truth the airship most probably drifted with the wind for the most part, but improvements followed, and despite setbacks the fourth ship, LZ4, set off from Friedrichafen on a 24hr proving flight for the government in August 1908. An engine breakdown forced an unscheduled descent near the small town of Echterdingen, when a sudden squall blew the ship into the trees and it was destroyed by a hydrogen fire. Count Zeppelin believed this disaster to be the end for his airships, but the German people thought differently. In what became known as the 'miracle' of Echterdingen they rallied behind his cause, transforming the seventy-year-old count to the status of national hero. Donations began to pour in, enabling Zeppelin to continue

The wreckage of the LZ4 at Echterdingen. This accident proved to be a turning point for the fortunes of the Zeppelin airship.

his work, and some of this money was used to form the Zeppelin Company and build a new construction plant on dry land at the edge of Friedrichshafen.

As a military man through and through, Count Zeppelin was gently relieved from the reins of power to become more of a symbolic figurehead, a talisman for nation pride, leaving others to direct the company's commercial efforts. Under the guidance of business manager Alfred Colsman the *Deutsche Luftschiffahrts Aktien Gesellschaft* (DELAG – which roughly translates as the German Airship Share or Holding Company) was established in 1909 to promote airship travel between major cities within Germany. The following year the LZ7, christened *Deutschland*, made its

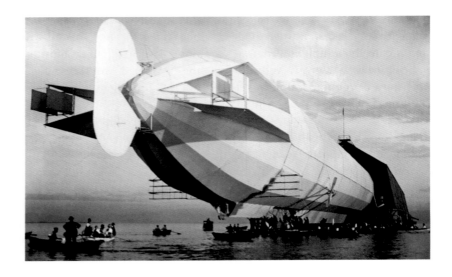

The LZ6 leaves the floating Zeppelin shed on Lake Constance.

inaugural flight, but was lost when it crash-landed in severe winds. Further bad luck followed; LZ6 burned in its shed and the LZ8 was smashed against the hangar doors by strong crosswinds. In command on that occasion was Hugo Eckener, a former journalist who had been won over to Count Zeppelin's cause and who would go on to become an instrumental figure in the future of the Zeppelin company. Eckener instigated a number of operational improvements, and by 1911, when the LZ10 *Schwaben* made its first flight, DELAG was ready to embark on a period of regular passenger flights. Three more ships followed over the next two years, the *Viktoria Luise* (named after Kaiser Wilhelm's daughter), the *Hansa* and the *Sachsen*. In the years leading up to the First World War these vessels made a total of 1,588 flights and carried 10,197 fare-paying passengers without a single injury.

Mapping the aerial oceans

Even Count Zeppelin began to turn his thoughts to the wider applications for his airships, including the prospect of transatlantic passenger services. Others were also inspired by Zeppelin's modest successes and fuelled by a spate of adventure novels with titles such as *The Great Airship* or *The Cruise of the 'Flying-Fish'*. Faith in the future of regular transatlantic voyages knew no bounds in the public's imagination.

ABOVE RIGHT: *A poster advertising DELAG's pre-First World War passenger services using the LZ11 Viktoria Luise. (US Library of Congress)*

RIGHT: *'The first passenger-carrying Zeppelin air-train,' proclaimed* The Graphic's *caption for this cutaway illustration published in 1910.*

In early 1910 *The Century Magazine* laid out the framework of such travels in a seventeen-page article entitled 'Over Sea by Air-Ship'. Accompanied by an illustration of an imaginary airship of 1915, just five years in the future, this revealed the 'surprising progress of German plans for a transatlantic service'. But, reading between the lines, was another and altogether more militaristic motive being hinted at? 'Behind all its thoughts and effort to conquer new spheres of activity, there is in the German Empire a determined purpose to achieve complete mastery of the air. Each test of successive air-ships has marked a step nearer that goal ...'

The main thrust of the author's case was that the mapping-out of the air currents was only a matter of time, and once this was completed airships would travel an aerial ocean in much the same way as ocean-going vessels plied the seas. This concept was nothing new. As early as 1816 Sir George Cayley had described the air as 'an uninterrupted navigable ocean, that comes to the threshold of every man's door.' Unfortunately the 1910 article's veneer of technological babble seemed to lend the theory an unwarranted air of scientific understanding. Thanks to the marvel of

wireless telegraphy, the article stated, 'all air currents in Germany have been charted for navigation.'

> Today, the telegraphic range equals an ocean liner's. From a system of wireless stations, hourly advices of the winds are available by all Imperial air-ships. Captains even now exchange the 'latest wind', independent of bulletins from stations. On this information, they go with the wind in one fourth the time lost in running against the current. This means that within the coming year, air-ships with passengers and mail may cross the ocean in one-and-a-half to two days.

Within the coming year? In justifying their optimism the authors referred to the recent performances of the Zeppelins, the emergence of the rival Schütte-Lanz airship company and to several imminent transatlantic attempts. It was taken for granted that airships would soon be capable of 'flying three to five thousand miles, to float for seven days, and to carry cargoes of five to seven tons, at a height of two to three miles above the Earth's gusts.' In reality these performance figures were, to say the least, ambitious. The greatest flaw in their argument was the belief that the air currents could be mapped so accurately. Yes, there are prevailing winds, but the aerial currents are not like the trade winds experienced at sea level, and the technology to fathom the vagaries of the weather in any detail would belong to the space age. More important than attempting to map the winds was the understanding of how the weather works, and in particular how the wind behaves around high- and low-pressure areas or near a frontal system. Eckener and his airship commanders of the 1920s and 1930s came to understand this, and they devel-

A contemporary German postcard of the LZ13 Hansa in flight.

Interior of the Hansa's cabin. Later pressed into service by the German Army, this airship was dismantled in the summer of 1916. (Dan Grossman)

oped a feel, an instinct if you like, for every minute clue to the weather's constantly changing patterns.

Back in 1910, and with the benefit of only ten years of airship flight, the meteorologists had barely dipped their toes in the aerial oceans. Professor Hergesell had undertaken extensive balloon and kite soundings over the 'trade region' of the North Atlantic on behalf of the German Admiralty, and from these he had mapped out several trade-wind routes. From Cologne, by way of Belle Isle, France, to the Azores and sweeping up to Florida, a distance of 3,700 miles (5,954km), was the shortest and most favoured, while alternatives went further south via either the Bay of Biscay or round the Spanish Peninsula to Madeira or Tenerife and on to either Peurto Rico or the West Indies and up to Florida.

At 40mph (64.4km/h), airships would, it was suggested, cross the Atlantic with the trade winds in less than two days, and the longer route with a stopover for fuel at Madiera would take less than four days. Given the faster west-to-east winds on the return leg, it was anticipated that airships travelling from America to Europe would complete the crossing in only one and a half days. 'For economic reasons, therefore, the trade-wind and the upper planetary drift are the natural navigable wind-rivers between Europe and America... Germany has stopped fighting the winds and means to put them to work. Marshalling their forces means faster and cheaper traffic.'

To support this aerial traffic, new rules of the air were to be devised, improved 'aero-time' maps were to be drawn, and aerial harbours would be established where air-pilots could safely take their craft into port. The new wind-rivers would also require new airships, as outlined in the article's vision for 1915:

From the standard of present development the air-ship of 1915 may be conceived as having a hull of rigid construction, one thousand feet long and eighty feet beam, with accommodation for one hundred and fifty passengers, and a crew of fifty-two men. The new air-liner will resemble a submarine, or rather a flying fish. All its parts will be compactly built into the hull.

In many ways the authors supplied design features that would become positively *de riguer* for all future schemes for aerial leviathans. A glazed prow as an observatory and ship's bridge, passenger accommodation including state rooms, central lounging, reading and dining saloons. Plus, best of all, an open-air observation deck on the top of the hull, complete with radio masts, kite-winch and 'boats' in the form of two small swift aeroplane scouts with ample space for launching and alighting. A veritable flying city! Some schemes being touted at this time were even more fanciful, such as the airship devised by engineers Radinger and Wagner. With a rigid hull of hollow paper tubes and steel bracing, it was to be 30 per cent lighter than Zeppelin's aluminium ships, a saving which would be converted into passenger and cargo capacity, and could fly for fifty days without any replenishment of the gas.

> Drum-shaped hollow compartments are to hold the sustaining hydrogen, none of which is lost through expansion by the sun, as any surplus will be compressed by automatic pumps into the hollow tubes... Engines of 242 combined horse-power are expected to develop a speed of forty to fifty miles an hour. This type of ship, soon to be placed in the construction cradle, is expected to cross the ocean easily with fifteen passengers.

A German rival to the Zeppelins

Back in the real world, much to his chagrin Count Zeppelin was not without a rival in the field of rigid airship construction within Germany. The LZ4's accident at Echterdingen had alerted a 35-year-old naval-architect named Johann Schütte to some shortcomings in Zeppelin's designs. With his experience working on hull design for Germany's largest shipping company, North German Lloyd, and as a professor at the Danzig Technical High School, Schütte identified several areas for improvement. A more streamlined and aerodynamically efficient hull shape, as opposed to Zeppelin's blunter-ended cylinder. A double frame, much like the double hull of a ship, with a strong keel for greater strength. Engine efficiency increased by connecting all propellers directly to externally mounted engine pods, and the clutter of the empennage to be replaced with simplified control surfaces at the rear, again as with a ship.

Unfortunately Schütte's initial approaches to the Zeppelin company were dismissed out of hand, prompting him to go into the airship business himself. With financial backing from the industrialist Dr Karl Lanz, the Schütte-Lanz company was formed in April 1909, with a factory established at Mannheim. Its first airship, the SL-1, was started in September 1909, and was what we would now term a 'proof-of-concept' prototype to provide Schütte with the opportunity to refine his designs. The SL-1 was a big ship with 723,950cu ft (20,500cu m) of hydrogen contained within its seven internal gas cells. The most noteworthy difference to the Zeppelin approach was an innovative crossed-spiral or geodetic structure constructed of laminated wooden girders braced with wires to form a rigid cage. The use of wood was commonplace in the fledgling aeroplane industry at the time, but problems would be caused by the effect of damp upon the hull's glued joints. All subsequent SL ships followed the more conventional Zeppelin ring structure, although still made of wood. Not visible from the outside, internal shafts had been incorporated to carry discharged gas

This photograph of the Shorts factory shows the process of manufacturing gas cells. At bottom right the women are scraping and washing small sections of goldbeater's skin which is then mounted on a linen backing panel and stitched into sections. A completed gas cell is being test inflated and examined for leaks.

upwards and out of the hull, a far safer option than the existing practice of allowing the gas to find its own way out through the porous outer cover.

When the completed SL-1 made its first flight, on 17 October 1911, its curving lines were pleasing to the eye. The externally mounted engines and the simplified cruciform control fins, three at first and a fourth added later, created the impression that here was an airship that meant business. As they often say in engineering circles, if it looks right then it probably is right. The proof came in the flight trials, with an impressive airspeed of 44mph (70km/h). Between October 1911 and December of the following year SL-1 made fifty-three flights. The longest of these, on 6 December 1912, lasted nearly seventeen hours on a journey between Rhienan to Biesdorf near Berlin. There it was handed over to the Imperial German Army, which was grateful to have a more businesslike airship constructor than Zeppelin to deal with, and SL-1 continued in service until an accident in July 1913.

The SL-1 had taken the rigid airship forward in leaps and bounds, with innovations that would benefit all future rigid airship designs, including Zeppelin's. Johann Schütte's grounding in the shipping business had opened his eyes to the airship's wider capabilities. In 1912, while the SL-1 was still flying, he was interviewed by an American journalist during a visit to the USA, and by all accounts he gave the young man a hard time. With a cigar clenched between his teeth he rebuffed the opening question about whether he would bring an airship (note, definitely not a 'Zeppelin') to America.

For why? Why should we bring it over here – an airship? For the money? Why should we bring an airship here, simply to send her up for the amusement of the people to look at and go away and talk about it?

Clearly Schütte was a little touchy on the subject of money. He felt that the German nation had done more than enough to keep Count Zeppelin's work going, while he had to seek private backing. There was also more than a hint of a chip on his shoulder as the airship pioneer who stood 'in the shadow of the titan'. Asked about the future, he asserted that with three to five years' more development the airship will easily cross the Atlantic. 'There are no technical difficulties in the way of building an airship today, but study of the winds is still needed.' It was a bold vision from a man who had only built one airship so far. Meanwhile, two teams of transatlantic hopefuls were readying their airships for actual attempts.

CHAPTER 2

The Early Attempts

Talk is cheap. It is also a lot safer than drifting above the mid-Atlantic waves, far from help. So, while many experts continued to theorise about the possibility of a trans-atlantic crossing, only a few brave souls were actually pointing their airships towards the great divide and preparing to fly!

Walter Wellman's airship *America*

Enter Walter Wellman, a New York journalist and self-publicist with a taste for adventure; a man who was not averse to filling the newspaper columns with reports of his own daring exploits. Wellman's airship adventuring began in 1906 with an attempt to fly to the North Pole with the *America*, an airship built by his engineer Melvin Vaniman. The envelope was 165ft (50.3m) long and held 225,000cu ft (6,370cu m) of hydrogen. Slung beneath this was the backbone of the semi-rigid airship, a framework to carry engines, crew and provisions. A particular feature of *America* was the 'equilibrator', a leather tube which trailed behind the airship. Its purpose was to carry supplies, and in theory it would regulate the airship's height above the ground by functioning in much the same way as the trail rope of a gas balloon. As the airship descended, more of the equilibrator's weight would rest on the ice, and thus with the load

LEFT: 'A forecast of the air-ship of 1915 – overtaking an Atlantic liner'.

The forgotten airship pioneer

Ernest Willows is not a widely recognised name outside the inner circle of airship aficionados. However, in November 1910 he made history by making the first flight from England to France over the English Channel, a modest achievement by later standards, but at the time a significant milestone in international airship travel. Willows' flight aboard his airship *City of Cardiff* was notable for being the first crossing by night and the first from England. It had not gone entirely without a hitch, however, and upon landing he was taken aback when French customs officials presented him with a bill for import duties on the hydrogen gas he had brought into France.

Willows was an extraordinary figure. Cardiff-born, he had built his first non-rigid airship at the age of 19, and he pioneered the method of tilting the angle of the propellers to direct their thrust. He attracted some interest in his airships from the army, but his exploits never earned him much money, and in 1926 he died in an accident while giving the paying public tethered rides in a gas balloon.

In November 1910 Welshman Ernest Willows made the first airship flight from England to France, but he was taken aback when the French customs officials demanded payment on the hydrogen gas he was 'importing'.

lightened the airship would rise. If it climbed too much then more of the weight was borne by the airship and it would descend.

The *America* had a theoretical range of over 2,000 miles (3,200km), but in the event it covered only 35 miles (56km) on the first bid for the Pole in 1906. So much for theory. When a second attempt was made the following year, a slightly enlarged *America* managed a measly 40 miles (64km). Evidently it was not up to the job, and when Admiral Robert E. Perry succeeded in reaching the Pole by dog-sled in April 1910, Wellman reluctantly gave up this quest and set his sights on the Atlantic instead. This was an incredibly bold leap into the unknown. A flight of around 3,000 miles (4,830km) at a time when the English Channel had been flown only a year earlier, by Louis Blériot in his Type XI monoplane. The first airship crossing of the Channel would not take place until October 1910, when the Clement-Bayard II took six hours to accomplish the 224 mile (360km) trip from Compiégne to Wormwood Scrubs in London. (British airship pioneer E.T. Willows made the first crossing from England to France the following month.) Unlike the

Walter Wellman poses on the deck of his polar expedition airship. (US Library of Congress)

meagre output was not nearly enough to push the bulky envelope all the way to Europe, and clearly a suitable tailwind would be a decisive factor if they were to succeed. And just in case they did not, there was the reassuring presence of a 27ft (8.2m) lifeboat slung beneath the keel, which would also serve as extra storage space in flight. Wellman still had faith in the equilibrator to regulate height, but this time it consisted of 300ft (92m) of steel cable to which were attached thirty cylinders, each 4ft (1.2m) long and 9 inches (24cm) in diameter, half containing fuel for the engines and the others empty to provide buoyancy.

The flight of the *America*

Incredibly, Wellman decided not to put the airship and its overgrown trail rope through any flight tests before the first assault on the Atlantic. It was a decision he would come to regret. The *America* was assembled and inflated within a wooden 'balloon house' at Atlantic City, New Jersey, and then began the wait for the weather they needed. Not too much wind for leaving the shelter of the hangar, but enough expected to get them all the way to Europe once launched. On 15 October 1910 conditions seemed perfect and, engulfed in swirls of autumnal mist and fog, the airship was walked out of the hangar by a team of 100 volunteers, many from the local fire department. Because of the sponsorship deals with various newspapers, media interest in the attempt was high, and the *Daily Telegraph* carried a report of the airship's departure:

channel flight, Wellman and his crew had very limited expectation of being rescued if things went pear-shaped mid-Atlantic.

Using some parts from the dismantled polar ship and combining them with a new French-built envelope, Vaniman put together what, in effect, was the *America No.3*. The 228ft (69.5m)-long envelope had a volume of 345,000cu ft (980cu m), and it is said Wellman had the hydrogen scented with oil of peppermint to aid in the detection of any leaks. Six internal air-filled ballonets maintained its shape. Once again the *America* was a semi-rigid, with an enclosed triangular keel running its length, 8ft (2.4m) wide at the top and narrowing at the base. Two engines, a four-cylinder Lorraine-Dietrich and an English-built vee-eight ENV, both nominally of 80hp, would drive the propellers. There was also a small 'donkey' engine for electrical power and to start the main engines. With hindsight this

One of the most adventurous journeys ever undertaken by brave men has just begun. The airship *America*, the second largest of her type in the world, has left Atlantic City and is now – so we earnestly hope – speed-

The crew of the America *huddle in the lifeboat as the RMS Trent comes alongside. (US Library of Congress)*

(250m) high, 430 miles (690km) east of Cape Hatteras.

Afterwards, Wellman reflected on their botched attempt: 'It was a trial worth making and covered over a thousand miles over rough seas.' Maybe, but for the most part his airship had behaved as little more than a free balloon, and in doing so had been at the mercy of the winds. There had never been much real hope of it reaching Europe, but the American press still looked kindly upon Wellman's courageous expedition. As the lead story in the *New York Times* stated: 'They had failed in their first attempt, but were not defeated … the Atlantic crossing was only a matter of another year or two, but it could be done, … a dirigible balloon could be built to do it, and … the only doubtful question was who finally should accomplish the miracle.'

The German press was altogether more scathing. 'At last the brutal, sensational and nerve-straining enterprise undertaken by an American journalist with the money of sensational American newspapers has come to an end.'

Germany's transatlantic hopefuls

The Germans were not prepared to relinquish their crown as masters of the air, and they had their own patriots who saw it as their duty to conquer the Atlantic. Josef Bruckner, a German-American journalist, proposed making the crossing by flying from Tenerife with the west-to-east southern hemisphere winds to the coast of South America. Bruckner persuaded the famous chocolate company Suchard to put up the money for a non-rigid airship, which was built by Riedinger of Augsburg, and in an early example of branded sponsorship this vessel was christened as the *Suchard* on 15 February 1910.

The *Suchard* measured 199ft (60m) in length and had a volume of 238,000cu ft (7,000cu m) although this was subsequently increased. A contemporary photograph reveals it to be an unremarkable craft with control surfaces both fore and aft. Instead of a conventional gondola it was equipped with a very businesslike wooden-hulled 'lifeboat' which housed the two engines, an NAG 110hp in tandem plus a 100hp Escher to drive the double-bladed propellers or the boat's marine propellers 'should it become necessary to come down on the surface of the ocean'. There was an additional 4hp auxiliary engine. A report in *Flight* for 26 January 1910 outlined the ambitious route in a surprising degree of detail: 'It is proposed to start from Cadiz then steer for Tenerife, from there to Puerto Rico, then generally beating back, sail over Havana, Cuba, New Orleans and so back to New York.

Despite the precaution of the lifeboat, concerns grew about the *Suchard*'s range and performance following a number of trial flights, and the Atlantic attempt was

'The airship Suchard of the Brucker transatlantic expedition.' A somewhat romanticized rendition of the airship's proposed crossing, published by Scientific American in January 1911.

postponed from early 1911 until the autumn to allow for further testing. Consequently the envelope was lengthened and the volume increased to 441,500cu ft (12,500cu

m). Improvements were also made to the control surfaces, and more flight trials were conducted. It is reported that the airship was taken to Tenerife in 1912, but it appears that the funds ran out before the Atlantic flight could take place. Another German airship team also proposed starting from Tenerife. Dr Gans-Fabrice of the Frankfurt Aeronautical Exposition planned mostly to drift with the wind to monitor the wind conditions on the route, but this project never made it off the drawing board.

Melvin Vaniman's *Akron*

In the USA the engineer Melvin Vaniman was back in the fray. Dissatisfied with the outcome of the *America*'s flight, he designed another semi-rigid to make a second attempt. A taciturn character once described by Walter Wellman as 'unemotional, unimaginative' and as a man 'living in his machines', Vaniman gained the support of Frank Sieberberg, the president of the Goodyear Tire and Rubber Company, to build the *Akron*. In appearance it was not that different from the *America*. With a

The Suchard airship was an early example of commercial sponsorship, although the branding is hard to make out in this photograph. Note the complicated control surfaces on the airship's bows.

slender 258ft (79m)-long envelope holding 400,000cu ft (1,130cu m) of hydrogen (approximately twice the size of the modern Goodyear blimps), it was the largest non-rigid of its time. The *Akron* had six propellers driven by two 110hp engines and one 80hp engine, plus a small donkey engine. There would be no equilibrator and to manage the variations of buoyancy caused by fuel consumption and temperature, Vaniman devised a method of switching the carburet-tor from the petrol fuel to the airship's hydrogen. If the ship was light they would burn hydrogen, if heavy, then more petrol, and by this means he hoped to avoid venting gas and conserve ballast.

The airship was completed in September 1911, but on its first test flight the following month the envelope leaked like a sieve, and Vaniman was forced to land. He then spent the winter repairing the airship, and on the second flight, in June 1912, disaster almost struck when one of the propellers became tangled in a rope, sending the *Akron* into a steep climb. Thankfully Vaniman's younger brother, Calvin, managed to crawl out to the propeller and sort out the broken blade.

Once again Atlantic City was chosen as the starting point for the transatlantic attempt, and at last, on the morning of 2 July 1912, the two brothers and three compan-ions clambered aboard *Akron* and prepared to head off for fame and glory. 'Let her go!' Vaniman shouted to the ground crew, and very gently the airship climbed away. Then, at about 500ft (150m) and heading out over the shore, it suddenly it began to ascend, and it is possible that Vaniman may have switched over to the hydrogen for fuel at this point. The crowds of onlookers gasped with horror when they saw a flash at the top of the envelope, and black smoke momentarily

Vaniman's airship, named Akron, emerges from the balloon house at Atlantic City, New Jersey. On 2 July 1912 it burst into flames shortly after starting a transatlantic attempt, killing all six crew members on board. (US Library of Congress)

obscured their view. A burst of flame split the envelope in two and, as if in slow motion, the *Akron* nosed over and began its fall to the sea. One body was seen to plummet from the gondola like a rag doll. All five men died that bright summer morning, in full sight of their inconsolable wives.

By the close of this first batch of trans-atlantic endeavours Walter Wellman had discovered that the aerial trade winds were always going to be fickle and difficult to predict. But he had lived to tell the tale, while Melvin Vaniman and his crew had paid the ultimate price for messing with hydrogen, and the German teams had never got off the ground. Now the sunny skies of peacetime were darkening as storm clouds of war gathered over Europe. The crucible of this conflict would accelerate aeronauti-cal development to such an extent that, by its conclusion, there would be a new breed of airships. In the interim, the civilized world was to discover the true impact of aerial warfare and the advent of total war.

CHAPTER 3

Going the Distance

When Louis Blériot succeeded in flying his monoplane across the English Channel in 1908, the author H.G. Wells observed that Britain was no longer 'an inaccessible island'. And when war was declared in August 1914 there could have been little doubt that Count Zeppelin's progeny, conceived to provide Germany with military domination of the air, would bring that war right to the doorsteps of Britain's civilian population.

The advent of total war

Having said that, Germany's airship forces, in the hands of the army and the navy, were ill-prepared at the outbreak of hostilities. DELAG's small fleet of three passenger airships was pressed into service, but the navy had lost two of its three airships in accidents. The army fared little better, with only seven rigids available, including the wooden-framed Schütte-Lanz SL-2. Initially

The forward gondola of a German Army Zeppelin, photographed in 1917.

the German High Command deployed its airships in support of the army at the front, to supply close-support bombing, but this policy was quickly amended when the army lost four of its rigids to ground-fire in quick succession. Use of the airships for strategic bombing was opposed by Kaiser Wilhelm, as he was reluctant to order attacks on non-military targets, and on London in particular because of his close blood ties with the British Royal Family. However, the airships' cause did have a dynamic advocate in Commander Peter Strasser, Chief of the German Naval Airship Division since 1913. Strasse had no qualms about the niceties of unleashing an all-out aerial war, and his views were shared by some members of the naval staff. 'We dare not leave any means untried of forcing England to her knees.'

On the night of 19 January 1915 the people of Great Yarmouth, Sheringham and the King's Lynn area were disturbed by an unfamiliar droning sound. Two German airships, diverted from their intended target of Hull by strong winds, dropped their loads of twenty-four 110lb (50kg) high-explosive bombs and incendiaries on the inhabitants of Norfolk. Four people died in the first air raids on British soil, and the attacks on London began four months later.

By May 1916 the German Navy had taken delivery of L30, the first of the radically improved 'super Zeppelins', which could carry a load of five tons of bombs. 'The performance of the big airships has reinforced my conviction that England can be overcome by airships,' wrote Strasser. 'The airships offer a certain means of victoriously ending the war.' But his optimism was to be shortlived, as the British perfected their defences against the raiders, with improved searchlights and anti-aircraft batteries surrounding London, and new aircraft which could climb to attack the

At the start of the First World War, in 1914, the Zeppelins represented an unimaginable threat with their ability to rain destruction upon civilian targets.

Zeppelins with incendiary bullets. These claimed their first kill on 7 June 1916, when Sub-Lieutenant Reginald Warneford, flying a Morane monoplane, dropped several bombs on the LZ37, returning across the North Sea from an aborted raid on London. The following year, on the night of 2 September 1917, Second Lieutenant William Leefe Robinson shot down the SL-11 over Cuffley, Hertfordshire. The fact that this was one of the Schütte-Lanz airships, built by Zeppelin's rival company, mattered little to the British, who exploited

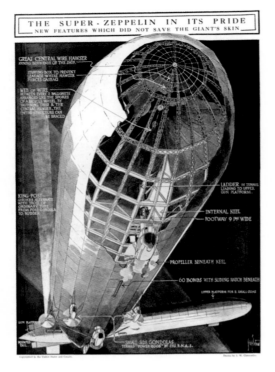

ABOVE: 'The Super-Zeppelin in its Pride.' This dramatic illustration, published in The Graphic in December 1916, reveals the inner workings of the German airships, and is said to be based on examination of the L33 which was brought down in Essex. Interestingly the caption comments that the type appears to be standardized, and 'includes the best points of the Schütte-Lanz airship'.

RIGHT: This detailed diagram from The Graphic of November 1916 portrays 'The Brain of the Zeppelin' and compares it with an example of the earlier open gondola or car.

the enormous propaganda value of the first 'Zeppelin' shot down over British soil.

The relentless war of attrition continued, pushing the Zeppelins to ever-higher altitudes to escape their attackers, and to achieve this they underwent enormous technological advances, from the first batch, the L3-class of 793,518cu ft (922,470cu m) to the first six-engined L30-class or Type-R 'super Zepps' of 2,000,000cu ft (55,200cu m), then up to the ultimate L70-class Type-X of 2,196,300cu ft (62,200cu m), which, in the case of L71, had a theoretical range of around 7,460 miles (12,000km) and a ceiling of 20,000ft (6,000m). For the Allies the capture of German airships, in particular the L33, brought down near the village of Little Wigborough, Essex in 1916, and the L49, forced down by French fighter aircraft near Bourbonne-Les-Baines to survive virtually intact in 1917, provided invaluable information for their own rigid airship designs. The American's ZR-1, the *Shenandoah*, was to be a virtual copy of the L49, while the L33 significantly influenced the design of the British R33 class, which included R34.

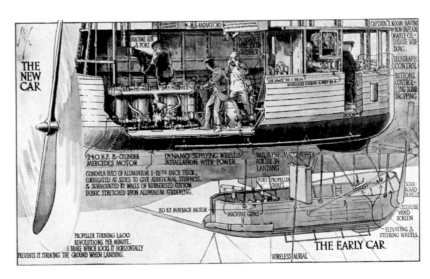

Setting new endurance records

With an increase in size and power came a corresponding increase in duration and range that took the Zeppelins to the point where they could have made an Atlantic crossing. This was amply demonstrated over the Baltic in 1917. The German Army airship LZ120 (the army's designation, which is not to be confused with the Zeppelin company's own LZ120 designation for the 1919 *Bodensee*) was an R-type with a gas capacity of 1,950,100cu ft (55,200cu m) and had six engines. Under the command of Ernst Lehmann it was deployed on patrol over the Baltic, usually for twenty-four hours at a time, on the lookout for enemy vessels, mines and to monitor merchant shipping. Lehmann recalls that during one of these long patrols the officers discussed the optimum length of such patrol flights:

> It was the consensus of opinion that it would be better, from a military viewpoint, to maintain a ship on patrol for much longer periods of time – for example, 100 hours or more. We also concluded that the length of flight might be limited solely by the petrol capacity of the ship... but other officers disagreed, for two reasons. They thought the motors would require overhauling before completion of such a long trip and further, that the men could not stand the exertion and strain of such extended periods.

Lehmann resolved to put the matter to the test, and with the LZ120 fully laden with bombs, water ballast, ammunition, equipment and provisions, plus a crew of twenty-nine, he calculated that there was sufficient lift remaining to carry 37,300lb

A recruiting poster celebrating Sub-Lieutenant Reginald Warneford's achievement in bringing down the LZ37 in June 1916. The realization that the airships could be successfully attacked was an important turning point for the British. (US Library of Congress)

When the L49 made a forced landing in woods near Bourbonne-les-Bains in October 1917, its crew failed to destroy the wreck, giving the Allies an important opportunity to examine a Zeppelin in detail.

(16,920kg) of fuel in the regular tanks and in thirty-seven additional temporary tanks hung up in the gangway. That would be enough for either fifty-six hours at cruising speed on all engines, or for more than 100 hours on reduced speed.

At a little before midnight on 26 July 1917 LZ120 lifted off from Seerappen, near Konigsberg, only to be held down alarmingly at first when the warmer air of a temperature inversion was encountered. The remainder of that first night was spent dodging thunderstorms over the open sea, but the rest of the mission patrolling the Baltic was largely uneventful. Of the six motors they used only three, sometimes four, at any one time. As

part of the exercise was to test how the crew would cope with longer patrol times, each man was relieved every eight hours during the first two days, and for the remainder of the flight they experimented with four-hour and then six-hour shifts.

> We cruised over the entire Baltic, in sunshine and rain, day and night. One of the motors ran so irregularly and with such vibration that the flange connecting bolts on the driving-shaft were nearly shorn off. Twice I had to stop and open the panel of the ship's hull to permit Chief Machinist Grozinger, and his comrade, Holzmann, to climb out on the propeller block. With nothing beneath them but empty space, they slid to the end of the shaft and put new screws in.

Provision was made for the boiling of water for drinks, and two cooking ranges heated by the engines were used to prepare hot food. The meals mostly consisted of little more than pea soup and poor-quality bread with some potatoes, pasta, a few hard-boiled eggs and some chocolate to break the monotony. Two tanks held 160 gallons of drinking water, and water for washing was drawn from the ballast. The only accommodation for the crewmen consisted of twenty hammocks distributed along the keel according to the airship's trimming requirements. Lehmann noted that for future long-duration flights the construction of sleeping quarters for the officers and bunk rooms for the crews would be desirable, along with a wash-room and dining room; all the basics of a passenger-carrier, in effect.

Lehmann described their extended time in the air as 'nirvana', and with little happening on the Baltic to demand their attention it was a welcome reprise from the stresses of the war. The control car was converted into

IT IS FAR BETTER TO FACE THE BULLETS THAN TO BE KILLED AT HOME BY A BOMB

JOIN THE ARMY AT ONCE & HELP TO STOP AN AIR RAID

GOD SAVE THE KING

The Zeppelin raids were designed to strike terror into the British, but instead they provided powerful material for the recruiting posters. (US Library of Congress)

for the Admiralty as copies of the German wartime Zeppelin L33 which had come down largely intact near Little Wigborough in Essex. Known affectionately as 'Tiny' by its crew, the R34 shared the Zeppelins' fine streamlining in its sleek 645ft (197m)-long profile. With a hydrogen capacity of just under 2,000,000cu ft (56,600cu m) it boasted a gross lift of 59 tons and was designed to cruise at 45mph (72km/h) with a possible maximum speed of 62mph (100km/h), although in practice the performance figures were slightly lower. The R34 was powered by five 250hp Sunbeam Maori engines, but the Admiralty's preference would have been for Rolls-Royce engines.

The R34 had arrived on the scene too late to carry out its intended wartime patrol duties, and it was the Air Ministry who decided to send the airship on the double transatlantic crossing. Initially the invitation had come from the American Aero Club and, ostensibly at least, it would be an important opportunity to study meteorological conditions over the North Atlantic and to demonstrate the ability of the rigid airship to undertake long-distance journeys. It was also an opportunity to show the Americans a thing or two, as they had only their fleet of little non-rigid blimps and nothing on the scale of the R34. The trip was also justified by the authorities as a means of forging 'a new link, by way of the air' between the two countries.

The senior officer on the transatlantic flight was Air Commodore E.M. Maitland; a quixotic figure with an unsurpassed enthusiasm for airships and a personal fascination with the parachute. He was the first person to parachute from an airship at a time when the parachute's value as a life-saver was much underrated generally. The captain of the airship was Major George Herbert Scott. Known to everyone as 'Scotty', he was the most skilful British airship pilot. In total the R34's crew numbered thirty men plus two carrier pigeons (it had been customary practice to carry these during the war, but what purpose they could serve out over the Atlantic was unclear) plus a lucky black cloth cat provided by the girls at Beardmore as a mascot.

Although Alcock and Brown had beaten them to the first non-stop flight two weeks earlier, the crew of R34 were to tackle the far more difficult east-to-west crossing against the prevailing winds, and then return to Britain to complete the first double crossing. Throughout this epic journey Maitland diligently recorded events in his personal log as he perched in odd corners of the airship. Such was the level of public interest in the flight that his account was

Provisioning the R34 with OXO and other foods previous to her trip across the Atlantic from East Fortune to New York.

OXO

OXO Ltd. Thames House, London, E C 4

Product placement is nothing new, as shown by this 1919 advertisement for Oxo featuring the R34.

published in 1920, and it provides a fascinating insight into life high above the Atlantic waves.

Departure

On the day of departure, 2 July 1919, the weather at the Airship Station at East Fortune, Scotland, was wet and breezy with the wind whistling round the corners of the big airship shed. At first sight these conditions looked far from suitable, but the weather reports suggested that by mid-Atlantic they should be more favourable. An area of low pressure was situated in the North Sea and moving slowly southward, while a high-pressure system dominated the greater part of the North Atlantic, with another reported over the Great Lakes of Canada. The result was unusually calm weather in mid-Atlantic, while the winds blowing over the west of Scotland and north of Ireland would hasten the airship's progress until well out to sea.

At 1.00am the officers and crew climbed aboard, dressed in their warm flying clothes and having had a hot dinner to fortify them on their long journey. Among them was a representative from the US Navy,

Lieutenant Commander Zachery Lansdowne, who would become a leading figure in the lighter-than-air scene in the USA, and Major E. Pritchard on behalf of the British Admiralty. The airship also carried a few small items of cargo, including a consignment of platinum for a firm of New York jewellers, and English newspapers for the editors of the *New York Times* and *Public Ledger*.

By 1.30am everything was ready, and a bugle sounded the signal for the handling party to commence walking the airship out of the floodlit shed stern first. With the wind blowing from the north-east and the shed facing south-west, R34 was swung round to face into wind as soon the tail was clear of the shelter provided by tall windscreens projecting outwards from the end of the shed. Held by a handling party consisting of nearly 700 people, mostly airmen and soldiers, the airship was at its most vulnerable exposed to the blustery conditions, so no time was lost in getting it airborne. At a signal from Major Scott the bugle gave the command to 'Let go' and R34 rose upwards to the cheers of the crowd, to be immediately engulfed by low clouds.

The R34 emerges from its shed. Note the size of the ground-handling party.

The voyage begins

As R34 headed northwest for the Clyde, Maitland described the experience as they moved off into the night:

> When flying at night, possibly on account of the darkness, there is always the feeling of utter loneliness directly one loses sight of the ground. We feel this loneliness very much tonight; possibly owing to the fact that we are bound for a totally unknown destination across the wide Atlantic. Such a feeling is only momentary, however, and is soon dispelled by the immediate need for action. Scott takes his ship up to the 1,500ft [458m] level, and lets go a ton of water ballast so as to clear all obstacles, the ship getting away about one ton heavy.

With a following wind of around 25mph (40km/h), R34 had a good ground speed estimated at 66mph (106km/h). The airship was fully laden with 4,900 gallons of petrol, and Scott wanted to keep it fairly low at this stage of the journey and would only go higher as fuel was consumed. But he also needed to negotiate the Scottish hills, which rise to 3,000ft (915m) in places. Continuing north-west, the airship entered further swathes of patchy cloud, and those on board got only an occasionally glimpse of the ground. Around 2.05am they spotted the twinkling lights of ships on the Firth of Forth. Passing over the Forth Bridge and Rosyth, they saw a trail of lights as an express train passed underneath, its glowing funnel spewing out a plume of white smoke.

Dawn comes early in Scotland in July, and by 2.30am the sky was already tinged with light as R34 followed the route of the Firth of Forth and the Clyde canal just south of the high Lennox Hills of Stirlingshire. The lights of Glasgow soon passed by, and

The R34 departs from East Fortune on 2 July 1919, rising into low cloud to the rousing cheers of the crowds. 'The coming of civil aeronautics,' announced The Sphere.

as the first rays of the sun tinted the sky salmon pink, the mountains of Loch Lomond came into view. It had been a long night, and Maitland headed off to try the sleeping arrangements. The ship had never been designed with passenger flights in mind, and the accommodation consisted of fifteen hammocks, to be shared by both watches, slung from girders along the triangular internal keel running from end to end of the airship. 'Getting in is quite an acrobatic feat, and falling out is better avoided in a service airship like R34, because there is only a thin outer cover on the underside of the keel on either side of the narrow walking way, and the luckless individual who tips out of his hammock would in all probability break through this fabric cover and soon find himself in the Atlantic.'

The riggers were accustomed to making

An illustration from The Sphere, August 1919, showing the main control cabin of the R34, with the captain facing the ballast chart and control levers on the left.

their way about the girders as they tended to the gas cells and inspected the ship, and the meteorological officer had a particularly precarious climb up the 100ft (30m) ladder to the top of the airship where he could obtain an unobstructed view of the night sky.

Five hours into the flight R34 entered a widespread area of murky fog and, combined with the weight of patchy rain falling on the envelope, these cold and damp con-

The view looking back from the control car of the R34 towards the rear engine pods as land gives way to the sea.

ditions were making the airship heavy, forcing Scott to keep her 12° up by the bow to maintain height. Over breakfast, served in the officers' living-room, separated from the crew's quarters by only a flimsy curtain, the officers discussed the merits of airships over surface vessels for transatlantic travel, and were unanimous in their agreement that even in high winds the rigid airship was far steadier. As Maitland noted, 'In the future, people who are bad sailors will prefer to make the long sea passages by airship if only to avoid sickness.'

As the morning progressed, the higher layer of cloud cleared to reveal a blue sky, while the lower cloud bank continued to shroud the sea below. Exposed to the warming influence of the sunlight the gasbags began swell to near-maximum capacity, and accordingly Scott took the ship down to 1,300ft (400m) and into the cloud bank to avoid super-heating. He could not afford to vent off hydrogen at this stage in their journey with the airship still heavily laden. They were making steady progress at 40mph (64km/h) and all but the two side engines were stopped to conserve fuel.

As the fog became thicker and thicker, Maitland's thoughts turned to lunch, which consisted of beef stew and potatoes, chocolate and water to drink. 'During meal times the ship is inclined to get an angle slightly down by the bow, owing to officers' and crew's dining quarters being situated too far forward. This must be corrected in future designs. Necessary to send some of the crew aft to correct trim. I notice they don't forget to take their food with them!'

Afterwards he decided to 'turn in', but his slumbers were soon disturbed when Scott came to his hammock with news that they had located a stowaway. Aircraftman William Ballantyne had been one of three men stood down for the transatlantic flight to make way for the VIPs, but, determined not to miss out on the adventure, he had hidden in the darkness high up on one of the longitudinal girders between the gasbags. It had been an unpleasant few hours for Ballantyne, who had come out of hiding because he was feeling ill from breathing in the hydrogen fumes. Stowaways, no matter how well intended their motives, could have endangered the entire flight, as the extra weight had not been allowed for in the

'Guiding' the R34 across the Atlantic by wireless from a room at the Air Ministry in the Hotel Cecil, London.

calculations. If he had been discovered over land there is no doubt that Maitland would have had him dropped off by parachute. But with the airship well out to sea that was no longer an option, and after being treated with quinine Ballantyne was allocated duties for the remainder of the flight. Later that afternoon a second stowaway was revealed in the shape of a tabby cat. Apparently she had been adopted by one of the crew a few weeks earlier and had

His Majesty's Airship R34 in flight over the Atlantic.

previously been aboard R34 on a 56hr endurance flight over the Baltic.

As temperatures fell in the evening the gasbags began contracting and Scott increased the airship's altitude to 2,000ft (600m) to clear the cloud bank. For a while the crew found themselves in an enchanting dream-like setting with an ocean of white extending to the horizon beneath a crystal blue sky. 'We feel in a world of our own up here amidst this dazzling array of snow-white clouds. No words can express the wonder, the grandeur, or the loneliness of it all...'

By the next morning, 3 July, they were over halfway to Newfoundland and already beyond the wireless range from East Fortune. The southeasterly wind was starting to pick up, and by mid-afternoon the sea had become rough.

Lunch consisted of cold roast beef with

The R34's stowaways, William Ballantyne and Wopsie the cat. Ballantyne's presence infuriated the airship's officers, especially when he became the darling of the New York press.

one cold potato each. 'We are short of potatoes,' Maitland lamented, 'having eaten too many yesterday!' But they still had bread, butter, cheese, chocolate and tea. If Maitland's interest in the meal arrangements seems endearing, it should be remembered that part of his role was to assess the airship's potential for future long-distance services and to suggest improvements. Food plays an important part in breaking the routine on a long flight, especially when there is little to see but clouds. However, the facilities on R34 were very limited and water could only be heated on a metal hotplate welded to the exhaust pipe of one of the engines.

That evening the thick fog was making life on the airship unpleasantly cold and there was an air of anxiety among the crew. They were now 350miles (560km) from St John's, Newfoundland, and they had already encountered heavy rain associated with a depression when fierce squalls began to pitch the airship up and down. 'The rain is driving through the roof of the fore car in many places, and there is a thin film of water over the chart table. The wind is roaring to such an extent that we have to shout to make ourselves heard... Time for evening meal, but no one gives it a thought while this entertainment is going on.'

Friday 4 July brought a wonderful sunrise and a welcome reprise from the buffeting of the night. The fog was still obscuring the sea, although an occasional break revealed conspicuously bluish-green patches of water and a number of large icebergs, causing Maitland to muse on the fact that the airship liner of the future will, at least, be immune from any iceberg risk. It was, after all, only seven years since the ocean liner *Titanic* had so famously struck one. Shortly before 1.00pm there was cause for celebration in the cramped control car.

Land in sight. Hooray! First spotted by Scott on starboard bow. A few small rocky islands visible for a second or two through the clouds and instantly swallowed up by them. Altered course S.W. to try and get a closer look at them. Eventually make them out to be north coastline of Newfoundland. This is quite the most thrilling moment of our voyage – great excitement on board. Whether or not we now succeed in getting through to New York, we have at any rate successfully accomplished the first stage of our adventure, and are the first to bridge the gulf from east to west by way of the air.

Arrival in America

Having reached dry land, the R34 still had to get to its final destination, Mineola Field on Long Island, New York. Flying against strong headwinds, the airship's progress was reduced to almost a crawl at times and, dodging violent thunderstorms, it was a rough ride as the nose pitched heavenward one moment and lurched downwards the next. By the end of the third day, worries over the fuel supply had become the main concern for Scott, who had to weigh up the options of carrying on and possibly running out of fuel, or making an intermediate stop for fuel at Montauk on the eastern tip of Long Island. The next morning the crew chief, Lieutenant Shotter, organized a party of men to inspect every fuel tank and, where appropriate, they drained the depleted tanks to collect every last drop. Scott calculated that they had just enough left to reach the mainland, but the last part of the journey proved to be a desperate struggle against the clock as the fuel levels dwindled ever lower.

Having reached the airfield early on the morning of 6 July, the first foreigner to arrive on American soil by air was Major Pritchard, who leapt from the R34 with one

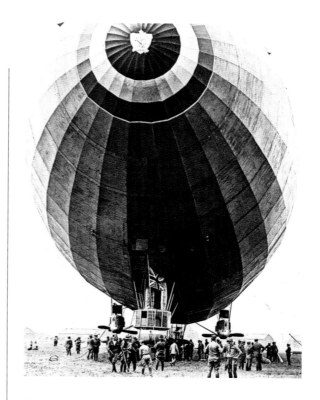

The R34 landed at Mineola, Long Island, on 6 July 1919 after a 108hr transatlantic crossing.

The R34's captain, George Herbert Scott, or 'Scotty', photographed shortly after the airship's arrival at Mineola. (US Library of Congress)

A children's picture 'scrap' of the R34.

of Maitland's beloved parachutes to take charge of the inexperienced American ground crew. The main handling party had not returned after being sent to Montauk, and in their place was a band of soldiers and sailors, none of whom had been near an airship before. At 9.54am local time R34 finally came to rest at the Hazelhurst Field, Mineola, accompanied by the rousing strains of 'God Save the King', played by a local brass band. Their journey across the Atlantic had taken 108hr 1min and had established a new world endurance record. It is estimated that they there was around 140gal of fuel left; enough for only two hours more. It had been a close call.

Over the next few days thousands of excited sightseers flocked to Hazelhurst Field to see this aerial marvel for themselves, and 500 military policemen were on hand to keep them under control. For the R34's officers and crew there was a chance for some very welcome sleep, not to mention the luxury of a hot bath. Most of the men were given 'shore leave' to enjoy the lavish hospitality of the New Yorkers, who enthusiastically wined and dined them. They even mobbed poor old Scotty and tore at his clothing for souvenirs, but much to the dismay of Maitland and his fellow officers it was William Ballantyne, the stowaway, who was to enjoy the greatest celebrity as a pack of news-hungry journalists followed his every move to milk his story to the full.

American soldiers served as ground handlers for the British airship during its stay in the USA. They are seen here surrounding the forward gondola.

'Aerotropolis' was a virtual city in one building. Taller than the Empire State Building, it includes landing strips on the roof and an airship port on one side. Colossal and totally impracticable, it was described in Popular Science *in 1939.*

all shapes and sizes, but especially gigantic. A 1929 article published in Germany revealed designs for Berlin's World Airship Airport, or as the Germans put it, *'Der Weltverkehrsluftschiffhafen'*. Written by H. Dominik and illustrated by A.B. Hanninger, this article demonstrated an appreciation of the problems associated with the ground handling of big airships. The airport was to have a central rotating shed that could be turned to keep it facing into the wind. An airship would dock at a high mast at the open end of the shed, and this would be lowered to ground level and retracted into the building with the airship attached. Mechanical mooring aids included a caterpillar tractor, or 'mule', which would hold

the aft of the ship via a system of chain ballast. Then, in true *Thunderbirds* style, the whole shed, airship and all, moved on rails to stand in line with similar sheds/airships awaiting their turn. Admittedly moving entire buildings as large as airship hangars would not be an easy task, although there had already been some examples of rotating sheds. The airship industry continues to explore schemes for mechanical mooring and handling to this day.

Taking this concept one massive and entirely improbable step further, the November 1939 edition of *Popular Science* featured Nicholas DeSantis's proposal for a 'Skyscraper Airport for the City of Tomorrow'. This entailed a 200-storey

Did airships really moor on the Empire State Building?

The notion of airships mooring to the top of the Empire State Building, once described as the 'looniest building scheme since the Tower of Babel', is the stuff of science fiction which has trickled down to become a staple feature within the murky landscape of popular urban myth. So did this ever actually happen?

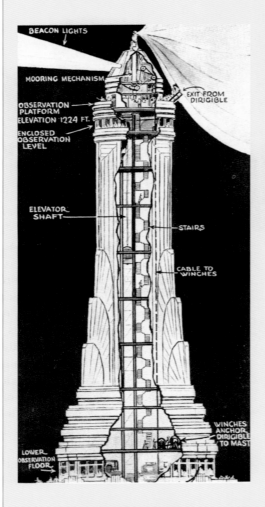

BEACON LIGHTS

MOORING MECHANISM

EXIT FROM DIRIGIBLE

OBSERVATION PLATFORM ELEVATION 1224 FT.

ENCLOSED OBSERVATION LEVEL

ELEVATOR SHAFT

STAIRS

CABLE TO WINCHES

WINCHES ANCHOR DIRIGIBLE TO MAST

LOWER OBSERVATION FLOOR

A cross-section of the Empire State Building's Zeppelin mooring mast.

As is so often the case, the facts behind this story are as strange as any fiction, as the accompanying illustration from an original brochure, *What Everyone Wants to Know About Empire State*, demonstrates. The story goes that the pair behind the erection of the building, financier John J. Raskob and former New York Governor Alfred Smith, decided to adorn it with a 200ft (61m) 'Zeppelin' mooring mast in order to convincingly trounce the Chrysler Building as New York's tallest building. Accordingly, by the time the Empire State Building opened, on 1 May 1931, its structure had been strengthened to withstand the pull of a moored airship, some of the winching equipment was installed and the eighty-sixth floor had been kitted out as a departure lounge for the would-be air travellers. But no one had really looked into the operational logistics of mooring an airship to a tall building in the middle of a city. For a start there was the hazard of approaching above the other buildings, such as the Chrysler Building, combined with the complex pattern of violent wind currents that swirl around such tall structures. Then there were the realities of getting the passengers on and off the airship on its quarter-mile high perch, not to mention the hazard of dropping tons of water ballast on the unsuspecting inhabitants of Manhattan. And one can barely imagine the outcome if the *Hindenburg* disaster had occurred in the middle of New York.

Smith had called upon the US Navy to dock the ZR3 *Los Angeles* — the former German-built Zeppelin LZ126 — on to the mast, but it declined and only sent a small blimp to hover nearby for the benefit of the newsreel cameras. Clearly the project was a non-starter, but it did endow the world's tallest building with its distinctive spire and two highly lucrative observation decks.

A visualization of a floating mooring buoy for trans-oceanic airships, published in Die Woche in 1930.

building around half a mile (800m) long and almost twice the height of the Empire State Building. On the flat roof there was a ten-lane landing area for aircraft, while a 'port' for dirigibles and blimps, complete with mooring beams to move the airships into the internal hangar spaces, jutted out from the side of the building. What this 'aerotrop-olis' lacked was any appreciation that airships needed to land and take off into wind, which is impossible with only a 180-degree arc accessible, or with any basis whatsoever in reality for that matter. But this sort of far-fetched pseudo-technology was the mainstay of such magazines, which variously featured airships that stayed aloft thanks to a mysterious lifting force gener-ated by 'gyradoscopes' and another using a vacuum cell. In fairness to DeSantis, it is quite likely that his inspiration came from genuine proposals to moor airships at the top of the Empire State Building.

Another German idea, this time for float-ing docking buoys, may have been inspired by the consideration being given about that time to creating mid-Atlantic floating bases or artificial islands for aircraft. The airship docking buoy was to be situated in calmer waters just offshore to act as a landing point from which passengers would be ferried between the airship and mainland by small boats. This has its parallels with the methods of operating flying boats. Its main advantage was that an airship would be able to rotate freely in any direction with the wind, but, inevitably, there were problems. Moored at low level, the airship could be in danger of striking the water with its tail, and while the practicalities of refuelling and so on could be overcome, the airship would have to return to a land base for routine maintenance and overhaul anyway.

All of these proposals, no matter how ludicrous, were grist to the mill for the airship's cause, as they kept the subject firmly in the public eye. Before we dismiss them out of hand, let us not forget that Count Zeppelin's visionary ideas had been greeted with widespread scepticism at first, and yet his progeny came to rule the skies. As Mark Twain once said, 'The person with a new idea is a crank, until the idea succeeds'.

CHAPTER 6

Disaster over the Humber

~ ~

During the last months of the First World War the British Admiralty issued a requirement for a new class of long-range rigid airship for maritime escort duties. In so doing they set in motion a course of events that culminated in Britain's worst airship accident, and one in which more people were to die than in the *Hindenburg*. Conceived in war, unwanted in peace, the R38 was supposed to have been the second airship to make the transatlantic crossing.

The Admiralty 'A' Class

The seeds of the R38's downfall can be traced directly to the Admiralty's over-ambitious demands for its new class of airship, known as the 'R38 class' or the 'Admiralty A Class'. These were to be capable of patrolling the North Sea for six days without support, as far as 300 miles (480km) from a home base, able to climb fast to an operational ceiling of 22,000ft

Almost completed, the R38 under construction at Cardington.

(6,700m), while carrying enough fuel for 65hr of flight at full speed, in addition to a veritable arsenal of weaponry including four 520lb bombs, eight 230lb bombs, a gun platform on the top of the envelope and a further twelve pairs of machine-guns.

The stipulation for the high-altitude capability had been inspired by the super Zeppelins of the First World War and their ability to climb out of reach of attackers. But what the Admiralty, and even the British airship designers for that matter, failed to grasp was that the Zeppelin height-climbers had been built on a solid foundation of technological expertise and the understanding that they were pushing the operational envelope right to the edge. The Zeppelin L70, which had come down off the Norfolk coast in August 1918, had provided the British with much information on the lightweight structure of the German high-achievers, but what it did not reveal was their inherent operational limitations. Instead the Admiralty wanted to copy the Zeppelin and make it better, and to achieve that the new airships would have to be bigger than any built so far.

With the outline design prepared by C.I.R. Campbell and his team at the Admiralty Airship Design Department, much of the detail work was produced by C.T.P. Lipscomb of Short Brothers, the company awarded the contract to construct the first of the 'A' class, the R38. Work began at Cardington, near Bedford, in February 1919, three months after the war's end. In September 1919 a further order was placed for three sister ships of the 'A' class, the R39, R40 and R41, but later this was cancelled as a peacetime economy measure, even though the first two had already been laid down by Armstrong Whitworth at Barlow. The immediate postwar years were a period of great uncertainty for those working

The Short Brothers works at Cardington at the time of the R38's construction. Note the single shed, while in the foreground are the company offices and the semi-circular layout of Shortstown.

within the airship industry, especially when the government announced that Cardington was to be nationalized under the Defence of the Realm Act and would in future be known as the Royal Airship Works. One consequence of the nationalization was that the now unwanted R38 was also to be cancelled. However, salvation was on hand from the other side of the Atlantic.

American interest

The US Navy had gained considerable experience with pressure airships, or blimps, and they wanted to extend their lighter-than-air capabilities with a large rigid. In July 1919 the Naval Appropriation Bill passed by Congress allocated funds for the building of a construction shed and procurement of two rigids, one to be built in the USA (the ZR1, christened USS *Shenandoah* in 1923) and the other sourced from overseas. The Americans wanted to get their hands on a genuine Zeppelin, but as most of these were deliberately scuttled by their German crews at the end of the war they missed out. Because the USA had not ratified the Treaty of Versailles, the remaining German airships were distributed among England, France, Italy, Belgium and Japan as part of the war reparations process. An unauthorized approach to the Zeppelin company to build a new airship for the US Navy was swiftly vetoed by the US Secretary of War.

With Germany's airship industry severely restricted in its activities by the Inter-Allied Control Commission, it was only natural that the Americans would turn to their former allies for a new airship, especially after the R34's successful double crossing, which seemed to confirm that the British had the know-how. When the R38 was offered to the US Navy, they readily agreed to pay the asking price of $2,000,000, which included training for the officers and crew.

Design of the R38

The size and shape of the new airship was dictated by its proposed performance requirements, suggesting a volume of around 3,000,000cu ft (84,900cu m) or more, but in the event the design had to be modified to fit the only available shed, and although it was still the biggest airship, the R38 evolved into a more modest 2,725,000cu ft (77,000cu m). The framework of girders that began to take shape at Cardington was 695ft (212m) long and had a diameter of 85.5ft (26m). To keep weight down the number of gas cells was reduced from sixteen to fourteen, and accordingly the bays housing them were longer and the distance between the main transverse ring frames was greater, with two intermediate frames instead of the usual single frame. The restrictions imposed by the limitations

of the shed space also necessitated the two mid-ship sections being straight sided instead of a more cigar-shaped profile, further robbing it of resistance to bending.

The R38 was fitted with six 350hp Sunbeam Cossack III engines to give a maximum speed of 71mph (114km/h), and had fifty petrol tanks for an endurance of approximately 144 hr, more than enough to undertake a delivery flight across the Atlantic. Two of the engine cars were mounted high on each side of the hull to keep the airship's overall height to a minimum, but this risked an increase in lateral bending moments during turning manoeuvres. The US Navy also required the airship to be fitted with mooring-mast gear which added about a ton in weight to the bows, which then had to be compensated with water ballast at the rear. Inevitably this modification caused a degree of 'hogging', the tendency to sag, which applied greater tension on the upper longitudinal girders and increased compression on the lower half. It all added up to a mishmash of changes and a multitude of complicated recalculations for the engineers, all without the benefit of modern computers.

Construction

Constructing the R38 was a bitter-sweet experience for the Cardington workforce. Each day brought probable redundancy ever closer. Yes, they took pride in their work, but the spectre of the abandoned framework of the R37, which stood beside R38 in the crowded shed, was a constant reminder of their vulnerability to the whims of the politicians. There has never been any suggestion that the men and women who built the R38 were dragging their feet, but none of them could have had much of an appetite for the job's rapid conclusion. Even so, pressure to complete was being applied by the British government and, increasingly, by the Americans.

In April 1920 the first of the US Naval Rigid Airship Detachment arrived at the airship base at Howden, Yorkshire, where the finished airship was to be based for the purpose of conducting a series of test flights before being handed over. With the delays in construction, training began on existing British airships including the plywood-framed R32, which was already slated for scrapping, and later the R33. The Americans had also been promised time on the R34, but in January 1921 this airship was damaged on a night flight after running into a hillside on the high Yorkshire Moors. Despite her crew coaxing the damaged airship back to Howden on two engines, gusty winds smashed R34's nose into the ground and by the morning it had been wrecked by the continual buffeting. Therefore the 'Howden detachment', as the Americans had become known, continued their training on the smaller R80, which had been designed by Barnes Wallis at Vickers but was also due for decommissioning. By the time the R38 was actually on its way to Howden many of the Americans had been stationed there for almost a year and a half.

Trial flights

In June 1921 the R38 was finally completed, and on the evening of 23 June it emerged from the shed at Cardington ready for its first flight. Registered as R38, the hull also bore the US Navy designation ZR2 in feint lettering, as can be made out in the construction photographs. The question of how long the proving trials should take, including familiarization flights for the Howden detachment, was a matter of debate between those responsible for approving its airworthiness, the Air Ministry and the US Navy. The RAF had appointed Major Pritchard, famous for his dramatic parachute arrival in the USA from the R34, to oversee the trial flights. The US Navy's Captain Maxfield was chomping at the bit to make the Atlantic crossing before August was out, but Pritchard was wary of releasing the ship before it had been properly tested, or of putting it into the hands of the American crew before they were ready. He prepared an extensive flying programme lasting 150hr, but the Air Ministry insisted that it should take no more than fifty.

The pilot in charge on that first flight from Cardington was Flight Lieutenant Archibald Wann. An accomplished airship

An engine car of the R38, fitted with a 350hp Sunbeam Cossack II engine.

The R38 in flight, wearing its American colours as the ZR2. (US Library of Congress)

pilot, he was outnumbered and outranked on this occasion by Air Commodore Sir Robert Brooke-Popham, Air Commodore Maitland as Airship Director, and the Air Officer Commanding Howden, who were accompanied by Commander Maxfield and Lieutenant Commander Bieg of the US Navy. At 9.52pm the ground crew were ordered 'Hands off!', and with the engines at slow ahead the airship was taken to 'pressure height' at 2,300ft (700m). (This is the point at which the gas cells are fully extended in the reduced atmospheric pressure.) Wann took R38 up another 200ft (60m) to test the automatic valves designed to prevent the cells rupturing. After seven hours in the air the R38 descended back at Cardington in the first light of dawn.

On the second flight, on the evening of 28 June, modifications to the control linkages were tested and it was found that the control planes were overbalanced. The airship was returned to the shed, where the rudders and elevators were cut back to reduce the control surface area by an estimated 10 per cent. Other last-minute changes included a doubling of the bracing on the frames supporting the heavy 2,000lb (900kg) water ballast bags, and a redistribution of the food lockers and water tanks in the crew's quarters to spread their weight more evenly. Clearly there were con-

cerns about concentrations of weight on the delicate framework, and a list of instructions was issued to the crew, including one stating: 'Not more than fifteen persons should be allowed in the crew space at any time'.

At last, on 17 July 1921, the R38 departed from Cardington to make its way to Howden. It was a chance to put the airship through its paces, and the airspeed was increased to 50kts (58mph or 93km/h) at an altitude of 2,200ft (670m). One of the American airmen had the elevator wheel when the controls suddenly became 'awkward to handle' and the ship plummeted in a 500ft (150m) dive before Lieutenant Pritchard brought its nose back up again. A hasty inspection with torches revealed that two girders had been damaged, and once at Howden the airship was taken into the shed for a more detailed examination, which resulted in the replacement of several girders and the strengthening of others.

Disaster strikes

By 23 August the R38 was ready for its fourth trial flight. After that, all being well, it would be handed over to the US Navy. It was proposed that it should be flown down to Pulham, where it would be moored at the mast and made ready for the delivery flight across the Atlantic. Unfortunately Pulham was concealed beneath a blanket of low fog, and the airship was taken out to sea to continue tests overnight. Pulham was still fogbound the following morning, and Captain Wann reluctantly decided to head back to Howden, performing further flight tests on the way, including speed runs at up to 60kts (70mph or 110km/h) with no untoward problems reported.

At 5.00pm a message was sent to the Air Ministry, advising that the airship was expected to land at Howden at 7.30pm. According to Tom Jamison, author of *Icarus over the Humber: The Last Flight of Airship R38/ZR2* and the recognized authority on the subject, there was a discussion in the cramped control car about the airship's ability to withstand rough weather as might be encountered over the Atlantic. The question was whether extreme control movements in the denser low-altitude air could simulate the stresses incurred in severe weather. Jamison believes that this is why a series of increasingly extreme rudder movements was instigated as the airship passed over the city of Hull and the River Humber.

Approaching from the northeast, the airship pushed through a scattering of cloud as it headed over the docks. Thousands of people were out enjoying the warm summer evening, and they had a grand-stand view of the unfolding drama. Several reported that the airship was making sudden changes in direction and pitch when a crease appeared, running diagonally across the hull. A cloud of water ballast vaporised into a grey mist as the R38 split open like an egg, sending men and equipment tumbling through the sky. The nose dipped, splitting away and then falling as the first explosion sent a shock wave that broke shop windows in the town. At least two parachutes were seen to fall as the burning nose section tumbled towards the Humber in plain view of the crowds on the Victoria Pier. There was another explosion as the front section struck the water, igniting the fuel into a sea of fire that engulfed wreckage and survivors alike.

The aft section, which did not catch fire, tilted and then descended at a more leisurely pace, following the nose in a slow drift towards the distraught onlookers. Four men were on board as it made a miraculous

This commemorative postcard depicts the moment when the R38/ZR2 broke in two over the Humber.

AIRSHIP R 38 WRECKED BY EXPLOSION OVER HULL AUG 24 1921

descent on to a sandbank in water which was no more than 3 or 4ft (1m) deep. Of those who had been on the airship, sixteen of the seventeen Americans died, along with twenty-eight of the thirty-two British crew. Among the victims were some of the leading figures in the airship world, including Air Commodore Maitland, Flight Lieutenant Pritchard and C.I.R Campbell, the airship's designer, who had become manager at the Royal Airship Works in 1920. For the Americans the loss of so many experienced airship men at this stage of their rigid airship programme was a bitter blow. The morning after the disaster the *Chicago Daily Tribune* was one of the many American newspapers to inform the public of the disaster, its headline banner proclaiming: '16 YANKS DIE ON ZR2, 1 SAVED – 27 BRITISH VICTIMS AS BIG BLIMP BREAKS IN TWO'.

Only Five of Crew Survive as the Ship Falls. Hull, England, Aug. 24 – The airship ZR2, yesterday the highest achievement of man in aerial flight, tonight lies in tangles of charred metal along the river Humber within sight of Hull. Of the forty-nine men, the pick of the world's greatest balloonists, who ascended in it yesterday morning at Howden, only five are alive in Hull hospitals this midnight.

After listing the sixteen American casualties, including Commander Louis Maxfield and Lieutenant Commander Valentine Bieg, the newspaper wasted no time in identifying the cause of the tragedy as 'defects in the structure due to saving weight'.

As with many such disasters, there were stories of miraculous escapes. Such as that of Ernest Davies, picked up by one of the many small boats which rushed to the tail section. He was landed at Victoria Pier and insisted on walking off the boat to the

cheers of the crowds. Another rescued from the tail was British airman Flight Sergeant Walter Potter, who stayed on with the airship service only to meet his death aboard the R101 in 1930. Two men were pulled from the control car in the forward section, including Captain Wann, who survived his injuries, and the USN Lieutenant Charles Little, who died before reaching the pier. The sole American survivor was Rigger Norman Otto Walker, a last-minute replacement for another crewman.

Because the airship had not yet been handed over to the US Navy it was still manned mostly by the British. Several of the American crewmen had not been allocated to the flight, and one of these was the American Lieutenant Richard E. Byrd, who later went on to achieve fame for his Polar exploration flights with aeroplanes. Byrd was establishing something of a track record of missed transatlantic airship flights. In 1919 he had applied for command of the US Navy NC-3 flying boat on its crossing with two other 'Nancies' from Newfoundland to Europe via the Azores, but his request had been denied. He did get to fly in the NC-3 as far as Newfoundland, where he was offered a place on the C-5

Photographic postcards showing the wreckage of the R38/ZR2 were bought by a news-hungry public.

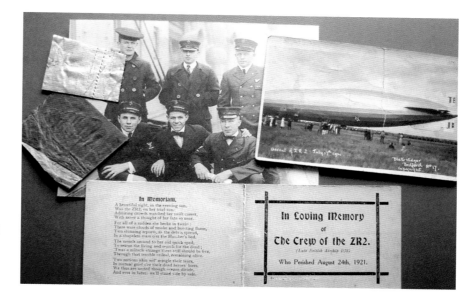

A selection of memorabilia from the R38/ZR2, including fragments of the gas cell and outer cover, plus a memorial card listing those who perished.

non-rigid airship which was due to accompany the three aircraft, but, as recounted earlier, this was lost on the eve of departure when it broke loose from its moorings.

The cause of the accident

The subsequent Committee of Enquiry into the R.38/ZR2 accident pinpointed structural weakness as the cause and concluded that no allowance had been made for aerodynamic stresses in the design process. No blame was apportioned by the enquiry, as this was not within its remit. Since then, several accounts have suggested that the R38 was subjected to extreme control inputs. Author Len Deighton wrote in *Airshipwreck*, 'No airship could take that sort of treatment. Even the steel hull of a destroyer would buckle if used in that way'. However, this grossly oversensationalises the situation, as those in command of the airship were highly experienced and competent airship men. Men who had weighed up the risks and decided to put the airship through a series of manoeuvres to make sure it was fit for purpose. Partly because the rough weather tests had not been possible so far, and partly because the

airship was expected to leave for America in the next few days. No stresses beyond those to be anticipated within the normal range of operation conditions had been applied, and the R38/ZR2 broke up because it was a weak ship, flawed in concept and design. This was an accident waiting to happen. If it had not happened over Hull, it would more than likely have done so mid-Atlantic, to become one of the many unexplained disappearances in the pioneering days of transatlantic flight.

The loss of the R38/ZR2 severely dented confidence in British airship design. It was certainly the worst possible timing for Commander Dennistoun Burney, who was pushing his Imperial Airship Scheme to link the far-flung corners of the British Empire. Undoubtedly the disaster left an enduring and indelible mark on the people of Hull, but beyond that immediate area it is a largely forgotten episode. History can be fickle, and had the newsreel cameras been there to catch the terrible last moments of the British and American crewmen it might have been a very different story. The R38/ZR2 was not the first airship disaster, and it certainly would not be the last.

CHAPTER 7

The *Amerikaschiff*

In the aftermath of the First World War, few in Germany had imagined how draconian the terms of the Armistice would prove to be, or to what extent they would manacle German industry for years to come. 'It was clear to us at Friedrichshafen that there

Dr Hugo Eckener, the legendary figure behind the Zeppelins in the interwar years.

would be no more military airships for Germany,' wrote Ernst Lehmann. With the ending of hostilities he had become assistant manager of the Zeppelin works, working under the indomitable Dr Hugo Eckener, although overall control of the Zeppelin organization was in the hands of Alfred Colsman, who had been with the company since 1908 as business manager. Now the organization was at its lowest ebb, and the management had the difficult task of keeping together its team of highly skilled workers and maintaining the company's undisputed lead in lighter-than-air when they were not allowed to build large airships.

Eckener was shrewd enough to realize that the restrictions being imposed by their former adversaries could not be enforced for ever, and in the interim he decided to construct two smaller airships to restart the DELAG passenger service within Germany itself. Surely the Allies could not object to that. The 796,350cu ft (22,550cu m) LZ120 *Bodensee* and the slightly bigger LZ121 *Nordstern* (*North Star*) were constructed from materials on hand and were completed in 1919. The *Bodensee* had even begun operations between Friedrichshafen

The route of the LZ126's one-way delivery flight to the USA, October 1924.

and their first lunch over France consisted of turtle soup, Hungarian goulash with peas and carrots, pudding and coffee. The men worked in shifts divided into two watches of four hours each, and as they progressed westwards the clocks were put back for each new time zone, the first occasion being as they crossed the Greenwich Meridian. Accordingly it was 1.00pm when the coast first came into sight at the mouth of the Gironde River. Lehmann summed up their feelings as the airship headed out across the sparkling waters. 'Our hearts beat faster. It was a solemn moment when the LZ126 left the European mainland to trust itself to the infinity of the Atlantic for the next few days. Before us, a west wind came up and blew against us. But one of the great advantages of airships is that they can choose their own routes.'

As the LZ126 crossed the Bay of Biscay they had a portent of things to come, with a stiffening southwest wind and rain clouds looming on the horizon. A new course was

set, parallel to the Spanish coast. It was dark by the time they passed Cape Vincent, the gusty conditions rocking the ship for several hours, and through the night the wind veered to the northwest, keeping the Zeppelin's flying speed to around 45mph (72.5km/h). With the steady consumption of fuel the airship was around 17,600lb (7,983kg) lighter, and they climbed to 5,000ft (1,500m) to vent some hydrogen. At a little past midnight Eckener radioed his position: 'We are over the open sea and making at present ninety miles an hour, making for the Azores. All the airship's machinery is in perfect order, working splendidly. All aboard are well and in excellent spirits.'

Shortly after noon on the Monday the LZ126 was over the Azores, and two bags of mail were dropped, all without stamps and relying on the generosity of the finder to mail the letters and cards. As they passed over the islands they saw people chasing after the airship as it disappeared into the layers of cloud. They were almost halfway to their destination.

Fifty hours into the flight the five engines of the LZ126 had consumed twenty tons of fuel, and ten remained. As a precaution the US Navy had ordered the cruisers *Detroit* and *Milwaukee* to stand by ready to provide assistance if called upon. On the crossing the ship passed low over several surface vessels, including the freighter *Robert Dollar*, which was near enough for the airship's crew to see the startled faces of the captain and helmsman on the bridge.

Strong breezes from the southwest retarded their progress and the weather pattern suggested that the airship would encounter strong headwinds as it continued on its proposed course towards New York. In the days before weather satellites supplied a constant stream of data on every square inch of the planet's surface, airship pilots such as Eckener had to construct their own weather pictures. Typical Atlantic weather systems consist of low-pressure areas encircled by isobars like a giant thumbprint on the charts. An airship on a transatlantic flight would not necessarily follow a straight line. Instead of fighting a headwind it was far better to steer around a weather system, riding the isobars to take best advantage of more favourable winds. The wind travels in a clockwise direction around a low-pressure system and anti-clockwise around a high-pressure system. Fortunately Eckener had an almost legendary understanding and feel for every nuance and clue offered by the ever-changing weather.

Whenever possible, meteorological data was obtained from surface vessels and ground stations, although the wireless communication was sometimes hampered by static. When a weather report from Cape West confirmed that the wind had increased to 35mph (56km/h) or more, they altered

This US Navy photograph shows the elevator-man's station in the control car of the LZ126/ZR3.

course to the northwest to seek more favourable easterly winds. A little later the *Detroit* reported southwest winds from its position near Cape Race, while the *Milwaukee* had more southerly winds 300 miles (485km) to the southeast. This data enabled Eckener to identify the position of a low-pressure system, and he immediately ordered a change of course to the northwest to take them around it, increasing the airship's speed to a more respectable 105mph (170km/h).

On the Tuesday morning the wind was backing from the east to the northeast and the characteristic fog banks off Newfoundland were engulfing the airship. They climbed to clearer skies at 5,000ft (1,500m), and then up to 6,650ft (2,000m) for a few hours, before descending back into the murk to ascertain the wind direction nearer the surface. The thick fog plunged the airship into semi-darkness and the thermometer fell sharply as bitingly cold storm winds battered them at 60mph (100km/h). The girders creaked and groaned under the strain until, after three hours, the storm finally abated.

During their third night aloft the crew of LZ126 could make out the blinking lights of Sable Island at the foot of Nova Scotia, and with only another 550 miles (885km) to New York, Lehmann recorded the sense of euphoria that swept over them. 'We had reached the American continent! We exchanged our marine maps for land maps, and tried to conceal our excitement from each other. But now we knew how Christopher Columbus and his crew had felt when, on the night of 12 October 1492, the lookout on the *Pinta* sighted land.'

Big in America

In the warm light of dawn the airwaves began to fill with messages of congratulations. At 4.15am the LZ126 droned over the inhabitants of Boston, and three hours later it had reached the man-made canyons and peaks of Manhattan. As they circled the Statue of Liberty the air was filled with the shrill whistles and sirens of the harbour and the factories, and thousands of New Yorkers poured out on to the streets to see the airship bathed in the dawn light.

The Germans were taken aback, indeed overawed, by the enthusiasm shown to them and their airship upon arrival in New York. The newspapermen tripped over their superlatives to sum up this aerial wonder they hailed as the 'Queen of the Skies', and the greatest accolade came from President Coolidge: 'It gives me and the people of the United States great pleasure that the friendly relations between Germany and America are reaffirmed, and that this giant airship has so happily introduced the first direct air-connection between the two nations.'

In Germany there were celebrations in the streets. Hearst newspaper journalist Karl von Wiegand wrote: 'Germany has rehabilitated itself in the eyes of the world'. When LZ126 touched down at the Naval Air Station at Lakehurst, New Jersey, on 15 October 1924, it became only the fourth aircraft to have crossed the Atlantic, and

At the conclusion of the 'Amerika-Fahrt' delivery flight on 15 October 1924 the LZ126 became the first Zeppelin to fly over Manhattan.

Having reached Lakehurst, New Jersey, on 15 October 1924, the LZ126 enters the shed for the first time. (US Library of Congress)

only the second after the R34 to make an east-west crossing.

The delivery flight had lasted for just over eighty-one hours, and now that the LZ126 had become the property of the US Navy this was one of the last times it would fly with hydrogen, as the Americans possessed the world's major reserves of helium. This is a totally inert gas which, although not as efficient a lifting gas as hydrogen, does not have hydrogen's deadly inclination to burn when mixed with air. Ironically, even the Americans could not supply sufficient quantities of helium to keep both the ZR3 *Los Angeles*, as the

LZ-126 was renamed in November 1924, and the ZR1 *Shenandoah* inflated at the same time, and the precious gas would be transferred from one airship to the other. The result of this tricky operation was that each airship was left idle for periods of extended 'overhaul', when in fact they just did not have the gas to fly. When the *Shenandoah* was destroyed in flight in 1925, ripped apart by powerful vertical air currents associated with squally conditions, the *Los Angeles* was forced to stand idle for seven months until new supplies of helium could be delivered.

The *Los Angeles* was to make a total of

The American public thronged to the airfield at Lakehurst to get a glimpse of the US Navy's latest acquisition, seen here before the application of its new markings.

Over the top of the world

Have you heard the one about the airship that flew from Europe to North America without crossing the Atlantic? It is a conundrum worthy of Lewis Carroll, but in May 1926 the N-1 did just that. This was a semi-rigid airship, one with a pressure envelope and a rigid keel, designed by Italian airship pioneer General Umberto Nobile. In 1925 he was contacted by Norwegian explorer Roald Amundsen, who planned to make the first flight over the polar ice cap. The N-1 was modified for the cold conditions, and as the expedition was being financed by the Aero Club of Norway it was renamed *Norge* (for Norway). Of medium size at 671,000cu ft (19,000cu m) and 348ft (106m) long, the airship was powered by three 780hp Maybach engines attached to the rigid keel.

Norge set off from Rome on 29 March 1926, and after stopping at Pulham, England, flew via Oslo and Leningrad to Kings Bay on the island of Spitzbergen, one of the most northernmost settlements in the world. With a team of seven-teen on board, including Nobile, Amundsen and the expedition sponsor, Lincoln Ellsworth, the *Norge* departed on 11 May. The following morning it passed over the North Pole, where the national flags of Norway, Italy and America were dropped on to the ice. By the third day the airship had reached the Eskimo village of Teller in North Alaska, where it was later dismantled, there being no plans to fly it back.

A second polar expedition was mounted by Nobile two years later, with the slightly larger N-4 *Italia*. This ended in disaster when the airship became heavy with ice and crashed, leaving nine men and Nobile's dog, Titina, stranded on the ice cap. This resulted in a protracted international rescue mission, during which Amundsen was killed when his aircraft disappeared on its way to assist in the search. Crushed by the ensuing widespread criticism for his role in the expedi-tion and its aftermath, the disgraced Nobile went on to continue his work with semi-rigid airships in the Soviet Union.

In May 1926 the N-1 Norge flew from Europe to North America without crossing the Atlantic. (US Library of Congress)

A contemporary postcard showing the LZ126, renamed ZR3 Los Angeles, moored at Ford Airport, Detroit.

ABOVE: *A rare view of the ZR3's rear gondola, with the Graf Zeppelin in the background, at Lakehurst in October 1928.*

BELOW: *A scrapbook picture of ZR3 Los Angeles.*

331 flights, accumulating some 4,180hr in the air, over an exemplary 14-year career. The longest-lived of any of the large rigid airships, it was finally withdrawn from service in 1932, after which time it was used only for mooring and structural tests. The US Navy had it broken up in October 1939 when its girders were cut into pieces and sold for scrap. Perhaps it was an ignoble end for the 'Queen of the Skies', but it had served its new masters well at a time when so many of its interwar contemporaries ended their careers in mangled heaps. Proof enough, if any were needed, that the Zeppelin company had built a very good ship indeed.

The flight of the LZ126 to America had been a positive turning point in allowing Zeppelin to continue building airships, and in 1926 the restrictions on Germany's involvement in international commercial aviation were lifted.

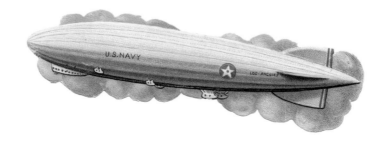

CHAPTER 8

Zeppelin Fever

~ ~

While the LZ126 had ensured the continuation of the Zeppelin company, the LZ127 *Graf Zeppelin* would become the most famous and successful airship the world has ever seen. It would prove beyond doubt that airships could conduct a long-distance service on a regular schedule, and it became the trailblazer for a breed of larger transatlantic-class airships still to come in the following decade. However, the Zeppelin company was not without rivals within Germany itself.

Schütte's postwar plans

Professor Johann Schütte had not completed an airship since the SL-21 in 1917,

and felt aggrieved at what he saw as the wholesale pirating of his design ideas by Zeppelin during the war, when the government had decreed that such niceties as patents should be put aside in the interests of the nation. The Schütte-Lanz airships had been noteworthy for their fine streamlining, simplified control surfaces with a single pair of rudders and elevators, the location of the engines in individual gondolas or pods and, initially at least, the use of laminated plywood as the main structural material. Towards the end of the war, however, by the time construction had started on the SL-23 and SL-24, Schütte had adopted tubular duralumin girders. The wooden structure had been scorned

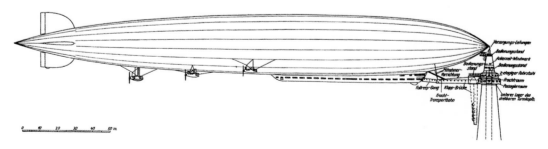

A Schütte-Lanz design of about 1920 for an 'Atlantic-type' passenger airship.

by the German Navy, which feared that it would be weakened in prolonged damp conditions. And those airships supplied to the army had been dogged by misfortunes, including a shed collapse at Leipzig and a big shed fire at Alhorn that destroyed the SL-20 along with four Zeppelins in January 1918. Worse still, several ships were completed and were either never commissioned or rejected by the intended customer.

Undeterred, Schütte continued to pursue his own aspirations for transatlantic passenger airships. As with the Zeppelin company, he looked overseas for backers to circumvent the postwar restrictions on airship construction in Germany. In 1919 he associated with Dutch partners to work on plans to build two 'Atlantic-type' airships, 750ft (228.5m) long with a volume of 3,590,000cu ft (101,700cu m). These featured an extended gondola for fifty passengers, which stretched beneath almost one-third of the forward hull. In America Schütte sought further support, although at Akron he discovered much to his dismay that a delegation from the Zeppelin company had been there before him. By this time he was openly feuding with Zeppelin, both publicly and in the courts. Matters became ever more complicated for Schütte, who severed the Dutch connection in order to pursue his working relationship with the curiously named American Investigation Corporation (AIC) on plans for a transatlantic service. In June 1922 the *New York Times* reported on the 'Advent of the American Air Liner':

> The intention is to make the first route a practical demonstration between New York and Chicago... The first airship will have a capacity of 4,000,000cu ft [113,200cu m] of helium and should cover the distance between New York and Chicago in ten hours' average time. About 100 passengers will be transported at night in berths similar to those on steamships. In addition to passengers, an airship of that size will be able to carry thirty tons of mail, express or other cargo... Coastlines or mountain ranges no longer are barriers requiring a change in transport methods. St Louis and Denver are as likely as any sea-coast cities to have direct airship communication with Europe, South America and Asia.

In *Airshipmen Businessmen and Politics 1890–1940*, author Henry Cord Meyer suggests that this article may have referred to Schütte's much larger design of 5,298,000cu ft (150,000cu m) which is sometimes referred to as the 'Pacific-type', although a transatlantic service was under consideration. By early 1924 it was becoming clear that the financial backing was not forthcoming for Schütte, and he parted company with AIC, which was left holding the US patents, resulting in Schütte fighting legal battles on both sides of the Atlantic. With the LZ126 on its way to America that same year he was rapidly losing the struggle for public recognition of his contribution to the design of the rigid airship.

The feuding with Zeppelin continued, but by 1928 Schütte's transatlantic ambitions had reached the end of the road. He had failed in a bid to design two new rigid airships for the US Navy, and to rub salt into his wounds his designs were rejected in favour of those submitted by the Goodyear-Zeppelin Corporation. (These were to become the ZR4 *Akron* and ZR5 Macon. See page 135.) In retrospect, historians will view Johann Schütte as the talented engineer whose significant contributions to the design of the rigid airship went largely unrecognized and unrewarded.

Rivalries and new partnerships

When the Zeppelin company was granted permission to go ahead with the LZ126, one of the original conditions had been for the sheds at Friedrichshafen to be demolished after its completion, a stipulation that can only be interpreted as a deliberate ploy by the Allied countries to scupper Zeppelin as a potential rival in the airship business. Fearing that they had no future in Europe, officials within the Zeppelin company began seeking partnerships overseas. At this time an American businessman and engineer named Harry Vissering entered

the scene. Vissering had become smitten by airships after a flight on the *Bodensee*, and in 1922 he privately published a volume in the USA entitled *Zeppelin – The Story of a Great Achievement*. This built upon Count Zeppelin's legacy and presented what was, in essence, a prospect for future Zeppelin transatlantic services. In it Vissering recorded the successes of the postwar DELAG ship *Bodensee* before expounding optimistically on the company's preparedness for the future: "The Zeppelin organization today is prepared to build, deliver and operate rigid airships for any purpose... It is now possible to produce quickly any type of commercial airship

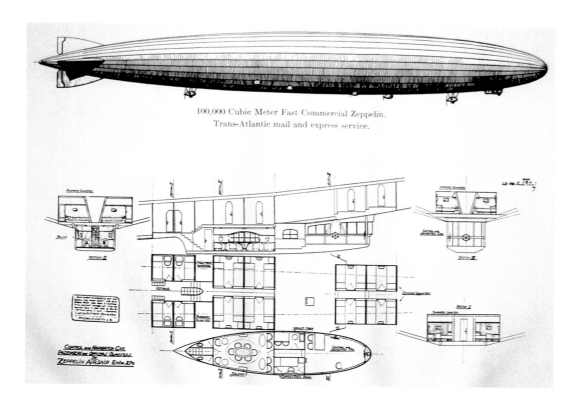

100,000 Cubic Meter Fast Commercial Zeppelin.
Trans-Atlantic mail and express service.

These drawings for a 3,532,000cu ft (100,000cu m) Fast Commercial Zeppelin were published by Harry Vissering in 1920, and feature passenger accommodation extended within the hull above the forward gondola.

from 700,000 to 7,000,000cu ft [20,000 to 2,000,000cu m] capacity."

Vissering envisioned five sizes of Zeppelins; a 'fast passenger' vessel based on the *Bodensee*, two intermediate sizes for training and medium distances and, most significantly, a 3,532,000cu ft (100,000cu m) transatlantic mail-carrier and an even bigger 4,768,000cu ft (135,000cu m) long-distance passenger Zeppelin. Illustrations of the airships showed the transatlantic mail carrier to be a conventional design outwardly, with the passenger quarters arranged in a forward gondola and the cabins situated above it, within the hull itself. The bigger passenger Zeppelin looked very much like a scaled-up version of the *Bodensee*, having an extended gondola with Pullman-style seating booths which converted into sleeping berths, and a saloon or drawing room towards the rear along with the kitchen and bathroom facilities at the back.

Specifics concerning a transatlantic operation remained sketchy, with only the generalities outlined by Vissering, which paralleled the existing Spanish proposals to a large extent:

> Carefully prepared calculations on some 600 flights made up and carried out from daily weather maps of the North Atlantic on methodically selected periods, have convinced the Zeppelin officials that a two and half day service could be maintained between Europe and America... With the exception of a few details, easily worked out in a brief experimental period, the Zeppelin organization could put such a service into operation at once, if permitted.

Vissering also referred to speculation regarding a New York-to-Chicago route, although he felt this might be more difficult

Vissering's 1920 visualization of the 'drawing room' area within a proposed transatlantic passenger Zeppelin.

to establish than the transatlantic service. Perhaps his most significant contribution to the transatlantic saga was his connection with the Goodyear Tire and Rubber Company of Akron, Ohio, and its president, Paul W. Litchfield. Vissering worked tirelessly at bringing the two airship companies together, and with the Spanish prospects fading an agreement was reached in 1924 for the establishment of the Goodyear-Zeppelin Corporation. This new company received the North American rights to Zeppelin's patents, along with a number of Zeppelin's key personnel, including chief designer Dr Karl Arnstein, who became an influential figure in the promotion of rigid airships within the USA and went on to design the *Akron* and *Macon*, the winner in the competition against Schütte.

The LZ127 *Graf Zeppelin*

By 1926 the restrictions previously imposed on civil aviation in German had finally been lifted by the Allies and the production facilities at Friedrichshafen were no longer under threat of demolition. Hugo Eckener was free to proceed with the construction a new transatlantic airship, the LZ127, provided, of course, he could raise the capital. From the outset it was clear that the Weimar government was more intent on building up its secret air force, and had no interest or money for airships. Recalling the enormous groundswell of support from the German people after the loss of the LZ4 at Echterdingen in 1908, Eckener established the *Zeppelin-Eckener Spende* (Zeppelin-Eckener Fund) to raise money for the new airship, and after two gruelling years of countless public appearances to promote the project he had raised 2,500,000 marks. Some of this came through what we now term as merchandising, with the sale of badges, postcards and so on. It was not enough to complete the airship, but enough to start construction. Later on the government was persuaded to contribute over a million marks to finish the LZ127.

A major limitation on the size of the new airship was imposed by the dimensions of the existing shed at Friedrichshafen. Consequently the LZ127 was narrower than was necessarily desirable, with a length-to-diameter ration of 7.8 to 1. Its overall length was 775ft (237m), and it had a maximum diameter of 100ft (30.5m). Nonetheless, the

This Zeppelin Eckener Spende embroidered badge was typical of items sold to the public as part of the fund-raising campaign for the construction of the LZ127.

gas volume of 3,707,550cu ft (105,000cu m) made it the largest airship built at that time. A major innovation was the use of '*Blaugas*', a gaseous mixture of propylene, methane, cetylene, butylene and hydrogen that was only very slightly heavier than air, which was used to fuel the airship's five engines, thus alleviating the complications of compensating for fuel consumed on a long voyage. This gas was contained in cells occupying the lower third of the hull, with a maximum volume of around 1,060,000cu ft (30,000cu m).

The airship's duralumin framework was mostly along established lines. There were seventeen main longitudinals, and the main circular frames were spaced at 49.2ft (15m) intervals, with two light intermediate frames in between. The main structural difference from the LZ126 was the inclusion of an axial gangway running the length of the ship, situated between the hydrogen and the Blaugas cells. The five engines, 550hp Maybach VL2s, were located in separate power cars arranged in two pairs on either side of the hull, plus a fifth one centred towards the rear. In operation the LZ127 proved to be faster than its predecessor by about 6mph (3km/h), although Eckener felt this needed to be improved upon for future North Atlantic services.

As with the LZ126, the passenger quarters were arranged within an external gondola at the front of the ship; 98.5ft (30m) long and 20ft (6m) wide to accommodate twenty people. At the nose of the gondola was the control room with its large outward

Assembling the girder work on one of the main rings of the LZ127.

One of the Graf Zeppelin's engine cars before installation of the 550hp Maybach engines.

sloping windows, control wheels for the rudder and elevators, plus gas valve and water ballast controls. Intersected with angled girders, it looked like a cross between a conventional ship's bridge and that of a futuristic spaceship. Situated behind the control room and running the full width of the gondola was the map room.

Behind that, the radio room was on the port side, with the electric-powered galley on the starboard. This tiny galley had a stove with two electric cookers, a water heater, a refrigerator and storage cupboards. Electricity was provided via small externally mounted wind-driven generators on either side of the gondola.

ABOVE: A close-up of the Graf's rear engine pod, which housed one of the five 550hp Maybachs.

LEFT: The gondola of the LZ127 during construction. This would contain the control cabin, radio and navigation rooms as well as the passenger accommodation.

Behind these operational areas was a corridor which also served as the entrance to the gondola, leading to a spacious lounge, or saloon, furnished with four tables, and this doubled as the dining area at meal times. Continuing backwards, another narrow corridor led to ten double-berth cabins for the passengers, five cabins on either side. At the rear of the gondola were wash rooms and toilets. By later standards the accommodation was cramped, but in comparison with all other forms of air travel it was positively luxurious. And while not on a par with the ocean liners in terms of space, it more than sufficed for the shorter journey time across the Atlantic. (In some respects the frequent airship passengers preferred the *Graf*'s cabins to those on the *Hindenburg*, most of which were without external windows.)

The *Graf*'s inaugural transatlantic flight

On 8 July 1928, the ninetieth anniversary of Count Zeppelin's birth, his daughter the Countess Hella von Brandenstein-Zeppelin christened LZ127 the *Graf Zeppelin*. Following further fitting-out the airship took to the air on 18 September for a 36hr endurance test flight. Such was the confidence in the new Zeppelin that after only five test flights, including one over the North Sea, it departed from Friedrichshafen on the inaugural passenger trip across the Atlantic on 11 October 1928. To help finance the first crossing, Eckener had sought revenue from a variety of sources. The main one had been the sale of press concessions for six newspaper reporters, additional revenue coming from fare-paying passengers and the carriage of special airmail and postal covers. Zeppelin had discovered very early on that the philatelists offered an enthusiastic and highly profitable market

This posed publicity photograph epitomizes the nation's pride in the new Zeppelin.

for the company's airship services, and many volumes have been published on this specialist aspect of the Zeppelin story.

Representing Hearst Newspapers were special correspondents Lady Grace Drummond Hay and her colleague Karl H. von Weigand, who were accompanied by photographer Robert Hartmann. Eckener shrewdly recognized the enormous significance of international press coverage to promote interest in the Zeppelin service, not just among the general public, but more importantly to sell the idea to the politicians and businessmen in the USA. To Lady Drummond Hay, an enthusiastic supporter of aviation in all its forms, fell the honour of being the first woman passenger to cross

the Atlantic by air. Together with Wiegand she recorded the day-by-day story of this remarkable voyage, and although they were frustrated in their attempts to radio lengthy reports back during the flight, the subsequent articles were widely syndicated and enthusiastically devoured by a Zepp-hungry public. But one should not be fooled by her gushy prose style and obsession with the finer points of her wardrobe. She was, after all, a product of the times in which she lived, and in her writing she was responding to the expectations of her readers. Underneath the glamorous exterior this lady was a tough cookie at heart, and her commentary provides us with a fascinating insight into the nature of transatlantic travel by Zeppelin, beginning with the luggage allowance:

Captain Hans von Schiller, who was in charge of such matters, gave out that the luggage allowance was to be 50lb (22.7kg).

ABOVE LEFT: The LZ127 Graf Zeppelin.

LEFT: The Graf Zeppelin at Friedrichshafen. It was arguably the most successful and famous airship the world has ever seen.

The interior of the Graf Zeppelin's control cabin, with the rudder-man's position at the front and the elevator controls on the left side.

It sounded little. It sounded a lot. All depends on what one wanted to take along. Very generously, he said we need not count in cameras and typewriters. Looking over my flimsy wearables, I felt very much elated. But shoes, that heavy beaded dress and new coat... They had to come out, wool-lined boots heavily made for the Friedrichshafen peasantry, had to go in, thick woollen stockings, sleeping socks, warm old jumper suit, woollen undies – and my quota was well exceeded. Compromising with a papier-maché grip, I eventually got the weight down.

Lady Drummond Hay's obsession with clothing almost caused her to miss the flight. Departure had already been delayed several days when the early-morning call came through to her hotel to get down to the airfield immediately. As she ran through the shed Eckener called laughingly, 'Hurry, hurry!'. She was already clambering on board when she suddenly realized that a coat specially designed for her by Gordon Selfridge had been left behind, and a porter was despatched in a taxi to get it. At last they were off, and there was time to describe the comforts of airship travel.

An airship, such as the *Graf Zeppelin*, is not only a wonder of the age but promises to prove one of the most useful wonders of civilisation. We will not arrive at our destination – in this instance America – in discomfort, with strained nerves, overtired or worn out. The passengers at least will step off as from a mighty, luxurious steamer, rested and prepared for a day's work or evening's entertainment... The journey began in earnest. Those who had not travelled on the Zeppelin before could not tear themselves away from the windows, running from one side to the other, exclaiming at every new phase of the scenery. Others were fascinated by the cabin arrangement, the charming little sleeping compartments, whose walls were no more than flowered chintz stretched from floor to ceiling... There is a luxurious

At mealtimes the Graf Zeppelin's *stylish saloon doubled as the dining room.*

A passenger cabin on the LZ127 Graf Zeppelin. *The back of the couch folds upwards to form an additional bunk bed.*

The compact galley area was equipped with a stove and two electric cookers.

*The Graf Zeppelin
leaves the airship shed
at Lakehurst, New Jersey.*

(11,260km) leg across Siberia to Japan, landing at Kasumigaura Naval Air Station in Tokyo three days later. The LZ127 then continued across the Pacific to Mines Field in Los Angeles, claiming the first non-stop transpacific crossing, before proceeding back to Lakehurst. The round-the-world flight had covered 19,500 miles (31,400km) from start to finish, or 30,830 miles (49,618km) if the initial and final flights between Friedrichshafen and Lakehurst are taken into account. It was a phenomenal achievement given that it was only three years after Lindbergh had made his perilous solo aeroplane flight from New York to Paris.

In May 1930 the *Graf Zeppelin* made its first trip to South America as the forerunner of a regular service on the southern transatlantic route. The forward march of the Zeppelins seemed unstoppable, and, as if to prove the point, in July 1930 the *Graf* was sent on a polar flight to rendezvous with the Soviet icebreaker *Malygin* deep within the Arctic Circle. Interest in the Zeppelins was at an all-time peak, both at home and abroad, and Eckener began to consolidate his plans for the next generation of transatlantic airships.

New horizons

In 1930 the International Zeppelin Transport Corporation (IZTC) was founded in the USA to study airship travel across the Atlantic. Colonel E.A. Deeds was appointed

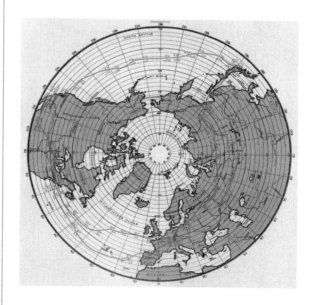

The route of the 1929 round-the-world flight, including the initial transatlantic crossing to Lakehurst, which was to be the official start point demanded by sponsor William Randolph Hearst.

A Hamburg-Amerika Line poster for the Graf's regular South America service to Rio de Janeiro, showing the aeroplane connection to Buenos Aires in the south.

chairman, Paul Litchfield acted as president and Hugo Eckener was on the board of directors. Meanwhile, the Zeppelin Transport Corporation (ZTC), instigated by three New York banking houses and with interest from several shipping and airline companies, would focus on the potential for airship operations between the Pacific coast of the USA and Honolulu, and explore the option of extending this to Manila and southeast Asia in the future. Confirming the cross-over between ZITC and ZTC, Paul Litchfield was appointed chairman of ZTC, and his vice-president at the Goodyear Zeppelin Corporation, Commander J.C. Hunsaker, was made its president.

Subject to suitable financial backing, the Goodyear-Zeppelin Corporation would build airships for both companies along the lines of Arnstein's designs for the US Navy's rigids, but incorporating high-specification accommodation for between forty and a hundred passengers. In addition to preparing designs and artist's impressions for the

The Graf Zeppelin in Brazil in 1934.

tentatively titled GZ-1, extensive dry runs were conducted to investigate actual transatlantic weather conditions on a series of hypothetical flights between Paris and Washington. From this it was established that, with a 58hr schedule in the summer and a 64hr schedule in winter for the eastward journey, and a 70-80hr schedule returning, passenger airships would reach their destinations ahead of time on 80 per cent of all journeys.

A report entitled 'Two Days to Europe in a Flying Hotel', published in the May 1930 edition of *The American Magazine*, was typical of the hype being generated around all this talk of international airship services:

> Man-made air monsters, as big as our greatest skyscrapers, will ride the ocean skyways like mighty silver-plated hotels magically floating on high. London will be within two dawns of Broadway... Paris and Berlin, Barcelona and Cairo will be way-stations on two-week vacation tours out of Cleveland and Kokomo, Pittsburgh and Hackensack. We shall skim the icy rim of the world in week-end sightseeing jaunts to the North Pole and roar around the earth in a week, amid the luxury of a Louis le Grand in his royal household.

With much of the flowery rhetoric inspired by the flights of the *Graf Zeppelin*, in particular with the circumnavigation, the journalist posed and answered the question of what might come next: 'Bigger airships! Picture a fourteen-storey building, more than three Fifth Avenue blocks in length, picking itself up and flying easily around the world in two weeks, the while it takes on and discharges passengers by airplanes

An artist's impression of future airship travel, published by Goodyear-Zeppelin in 1930.

without pausing in its flight. That is the sort of airship now *under construction*.'

It was certainly a beguiling image, and it was only natural that Goodyear-Zeppelin wanted to add commercial airships as an adjunct to its military contracts, as long as someone else put up the money. Ultimately, these plans did not come to fruition. This was largely due to bad timing, with the Depression in the USA and the loss of a number of rigid airships, including the R101 and both the *Akron* and *Macon*, undermining confidence in lighter-than-air transportation. By the mid-1930s the German-built Zeppelins had the field very much to themselves, with the redoubtable *Graf Zeppelin* continuing its regular transatlantic crossings and the bigger, as yet unnamed, LZ129 nearing completion at Friedrichshafen.

CHAPTER 9

Linking the Empire

∽◦ ◦∽

The Imperial Airship Scheme was a bold plan to link the British Empire's dominions by air. It grew out of a proposal put forward by Sir Dennistoun Burney of Vickers to build a fleet of six rigid airships to service the Empire routes, not an unreasonable concept considering the far-flung nature of the countries involved and the widely held view that only airships could operate on such long-distance routes. Ramsay MacDonald's newly elected Labour government thought so, and in 1924 they replaced

Number 1 shed, one of the two sheds still standing at the former Royal Airship Works at Cardington, near Bedford. The R101 was constructed in this imposing building.

Burney's plans with their own Imperial Airship Scheme. This put the emphasis on competitive designs for two 5,000,000cu ft (141,500cu m) rigid airships, each capable of carrying 100 passengers plus cargo over distances of up to 3,500 miles (5,600km). One would be built by the government at the Royal Airship Works, which now occupied the Shorts plant at Cardington, Bedford, and the other by a subsidiary of Vickers, the Airship Guarantee Company, in the large twin hangar at the former Royal Naval Air Station at Howden in Yorkshire. Thus was born the intense rivalry between the 'capitalist airship' R100 and the 'socialist' R101. Although both were considerably larger than any others of the period, including the *Graf Zeppelin*, they were only intended as the prototypes from which a new generation of bigger 8,000,000cu ft (226,400cu m) leviathans would follow.

In 1926 the Imperial Conference on the Future of Aerial Communications established the intended routes for the imperial airship fleet. The Royal Airship Works at Cardington would become the hub of the operations, with a number of routes reaching outwards like tendrils to the extensive red-coloured patches covering major parts of the world map. It was proposed that by 1930 there would be two main routes; across the Atlantic to Montreal in eastern Canada, and also southeast via Ismailia in Egypt and on to Karachi in northwestern India. Over the following five-year period these would be added to, the Karachi route being extended to Perth and Melbourne, in Australia, and on to Wellington in New Zealand. There would also be a new route around the coast of Africa, taking in Bathurst, Cape Town, Durban and Mombasa, and incorporating a direct link from Cape Town across the Indian Ocean to Australia. In preparation for the first phase, mooring masts were erected at Montreal, Ismailia and Karachi, the last-named location also having an airship hangar to serve as a secondary hub in central Asia.

Rival designs – the socialist and capitalist airships

In a deliberate bid to encourage innovation, the design of both R100 and R101 saw a move away from the established Zeppelin profile to a far more streamlined shape, like a fat cigar. In charge of the R100's design team was Barnes Wallis, who had considerable experience designing airships for Vickers, including the R80, one of the most aerodynamically shaped rigids to date. Assisting him as chief calculator in the days before computers was Nevil Shute Norway, who later became more widely known as the novelist Nevil Shute. In the light of the R38/ZR2's break-up over the Humber, greater emphasis was to be placed on the structural strength of both airships, and far more accurate calculations of

Registered G-FAAV, the R100 rides on Cardington's tall mast. Note that it had lost the tip of its pointed tailcone by the time this picture was taken. The mast was demolished during the Second World War, ostensibly to re-use the metal for the war effort.

the stresses and loading of the framework would be essential.

The R100 was designed with only thirteen longitudinal girders, which gave it its flat-sided appearance and made it much lighter than previous British rigids. Fewer longitudinal girders also meant there were larger unsupported areas of fabric outer cover, and this would prove to be an issue. The members of the triangular girders were formed as tubes from three helically-rolled and riveted duralumin strips. With an overall length of 709ft (216m) and a maximum diameter of 133ft (40.5m), the actual volume of R100 was slightly over the 5,000,000cu ft target at 5,156,000cu ft (146,000cu m), and it was designed to just fit into the hangar at Howden with the door height raised by 10ft (3m). The rear of the hull featured a distinctive tapering tail which, although aesthetically pleasing, did not withstand the aerodynamic pressures encountered during speed trials and had been replaced by a more rounded tail by the time it flew to Canada.

On the question of engines for the two ships, it had been originally intended that both R100 and R101 would have seven 550hp engines running on a new hydrogen-kerosene system, but as these could not be made available in time the decision was made to fit diesel engines instead. However, the R100 team rejected the diesels as being too heavy, and opted for six Rolls-Royce Condor aircraft engines, which ran on petrol. The consequence of this decision was that R100 might be deemed unsuitable for flight in the tropics, as petrol was considered to be more volatile than diesel in hotter conditions, and from an early stage it looked as if it could be confined to the north transatlantic route. Two of the R100's engines were fitted with reversing capabilities to allow for improved manoeuvring near the mooring mast. The engineers at Howden were amazed to discover not only that the rival ship, R101, would be equipped with the heavier diesels, but also that one of them would be rear facing, used only at the start and end of the flight. In fairness, the

Viewed from the rear, the R100 inside the shed at Cardington.

ABOVE: This advertisement for the pneumatic drills used in the construction of the R100, published in The Aeroplane in January 1930, is typical of the association that companies wanted to create with the prestigious airship projects.

RIGHT: The framework of R101 during construction; an incredible tracery of over 30,000ft (9,144m) of girders.

designers of the R101 had also wanted to fit petrol engines a few months before her first flight, but their request was denied by the Air Ministry. One of the penalties of working within a government team was the suppression of appropriate innovation in the face of official policy.

The whole capitalist/socialist slant led to an extraordinary situation in which the R101's design team knew about every aspect of the R100's construction, while the Howden team could only read of their rivals' work in the newspapers. Overall control of the production of both airships was under Wing Commander R.B.B. Colmore. Heading the R101 design team was Lieutenant Colonel Vincent Richmond, assisted by Squadron Leader E.M. Rope, with Harold Roxbee Cox (later Lord Kings Norton) as chief calculator. Their design incorporated a framework of rigid stainless-steel girders which were not braced with tensioning wires as with all

other rigid airships. Strength would be provided via deep reinforcing rings that actually limited the size of the gasbags. When, at a later stage, the thin wires holding the gas cells were let out to increase their volume, it created problems with chaffing on the framework.

Travelling in style

Passenger accommodation in R100 and R101 was grander and roomier than any yet seen in any airships, including that of the German-built *Graf Zeppelin*. The Airship Guarantee Company provided the following description of the R100's passenger area in a 1929 publicity brochure:

This structure is slung inside the hull of the ship and consists of three floors, of which the bottom is allocated to the crew and the two upper ones to the passengers. This

An interior view of the R100's passenger accommodation, which was arranged on two main decks.

coach is surrounded entirely by a double wall, through which air is circulated to obviate the danger of any inflammable gas or vapour penetrating to the living quarters. Cooking is carried out here in the electric kitchen. The passengers are quartered in two and four-berth cabins very similar to those found on a ship. Windows in the side provide the light and the view, and these are faced by promenades sufficiently spacious to accommodate a pleasant little dance.

This was the first time that the passenger areas on an airship had been located within the hull itself, and the double staircase and interior décor conveyed a homeliness and sense of British solidity; more Tunbridge

Wells teashop than the Bauhaus modernity typified by the later Zeppelins.

The R101's accommodation included a wide main lounge, decorated in white with gold inlay and curtains of Cambridge blue, with upholstered green wicker seats plus side couches and writing desks. This led out to the promenade deck, which had a decidedly nautical flavour, including some thoroughly British deck chairs. Sleeping arrangements in the fifty cabins were pretty basic, with simple bunk beds, a reading light and a stool for the luggage. The R101 also had an asbestos-lined smoking room on the lower deck which could seat twenty-four; this preceded the *Hindenburg*'s famous smoking room by several years.

It had been hoped that both R100 and R101 would be airborne before the Germans had the *Graf Zeppelin* flying, but by mid-1929 they were severely behind schedule and a year behind the Zeppelin. In the event the R101 was the first of the two to take to the air, on 14 October 1929. The R100 followed suit on 16 December, when it set

ABOVE: *The staircase linking the two passenger levels on the R100.*

ABOVE: *The R101's observation deck, complete with some thoroughly British deck chairs.*

RIGHT: *The spacious lounge or saloon area of the R101, fitted out with folding tables and lightweight cane chairs.*

course for Cardington, where it duly become government property. The R100 then underwent a series of six further test flights over the following seven months, including speed trials which demonstrated that it could attain a respectable 81mph (130km/h), making it the fastest airship in the world. It also made an acceptance flight of almost 54hr endurance and covering 2,050 miles (3,300km), flying over south-west England and the Channel Islands. One problem area the trials did highlight, however, was with the pre-doped cover, which at speed developed a rippling effect, like vertical lines or waves moving along the hull. Burney was quick to dismiss any concerns about this by pointing out that the contract had only called for a top speed of 70mph (110 km/h).

The R101 was also encountering its share of difficulties. During initial trial flights it displayed an alarming degree of instability, possibly as a result of the smaller fins failing to engage the airstream properly, compounded by the fact that it was simply too heavy. The problems began to mount one upon another; the gasbags developed numerous leaks, the gas valves had a tendency to open slightly when the ship rolled and the outer cover was weak in places. Drastic action was needed, and the R101 was returned to the shed over the summer of 1930 and sliced in two, just aft of the passenger accommodation, for a whole new section to be inserted at the point of maximum circumference. This increased its length to 777ft (237m) and the volume to 5,500,000cu ft (160,000cu m).

The big stretch. Over the summer of 1930 the R101 was taken into the shed and cut in two so that a whole new section could be inserted just behind the passenger accommodation.

A dramatic photograph of the R101's nose, showing the circular ventilation 'gill' holes and the lowered access hatch.

R100 flies to Canada

While the R101 was undergoing this drastic surgery it was left to the R100 to fulfil its contractual obligations by making a proving flight to Canada. At 2.48am on the morning of 29 July 1930 the R100 slipped its moorings at the Cardington mast. On board was a crew of thirty-seven. Major Scott was in charge of the flight, accompanied by a number of officials, including Wing Commander Colmore, director of Airship Development, and F.M. McWade, the resident inspector of the Aeronautical Inspection Department. Seven 'passengers' included Sir Dennistoun Burney and Nevil Shute Norway, who were on board to observe the airship's performance on behalf of her builders, the Airship Guarantee Company. This was not a commercial flight as such, but a test run for future transatlantic services, as outlined in an official press release:

> The flight of the Air Ministry airship R100 is being undertaken as part of the Ministry's development policy for airships with the

The gentleman of the press pore over one of the R101's engine cars inside the Cardington shed.

objective of testing out the reliability and behaviour of the airship on a long distance flight. Data will also be acquired which will be of value in deciding upon future policy with regard to the development of airships for commercial purposes, with particular reference to speeding up communications among the British Commonwealth of Nations.

Weather conditions were gusty and the cloud base down to around 2,000ft (600m) as the ship set off at a height of 1,000ft (300m) on a northwest course to Liverpool and then Malin Head, Ireland. From there the R100 would follow the great circle route south of Greenland, across Labrador and along the St Lawrence to Montreal. The airship was moving anticlockwise around the north of a depression, and the outer cover soaked up moisture from the low rain clouds and fog. At 11.00am the cloud dispersed and the airship's whale-like shadow could be traced as it flitted across the surface of the sea.

Lunch consisted of tomato soup, stewed beef, peas, potatoes, plus custard, beer, cheese and coffee. Enough rations were carried to feed forty-eight people for three days, plus two days in reserve and a further one-day supply of emergency rations. By mid-afternoon life on board had settled into a routine, and when not on watch the crew played cards or had an afternoon nap to compensate for their early departure. Compared with the flight of the R34 eleven years earlier, the facilities on board R100 were very comfortable, and as it was pleasantly warm there was no need for the insulated flying suits. Sixteen hours into the flight the airship was making about 51mph (81km/h) and it was anticipated they should make Newfoundland by the evening of the following day.

Radio communications were maintained with Britain initially before switching frequency for the Canadian Marconi long-wave station at Louisburg, Nova Scotia, but there was a period of nine hours mid-ocean when R100 was out of touch with both sides of the Atlantic. They only had contact with surface vessels, which passed on the news that a low-pressure weather system forming over Bermuda was moving eastwards. By early the next day the wind had veered and was increasing. Daylight revealed a layer of stratus cloud from horizon to horizon, but through a clear patch they spotted Cunard's *Ausonia*, their first sighting of a ship since leaving the coast of Scotland, and they exchanged greetings via the wireless and the *Ausonia*'s steam whistle. Running on four engines, the R100 continued to make steady progress. It was standard practice to stop each engine periodically so it could be inspected, but the only real concern on the outward trip was prompted by signs of leaking from some of the gasbags.

Further westward they encountered banks of fog and Scott took the R100 down to 700ft (200m) in an attempt to get underneath them. As the airship passed through the clammy mist it accumulated moisture, and small ridges on the outer cover funnelled 1.5 tons of water into the ballast bags located along the keel. When they returned to 1,500ft (460m) the glare of the sun bouncing off the fog was so bright that a curtain was rigged up in the control car to shield the steering coxwain's eyes. The thick cloud banks persisted, and to alleviate the tedium fresh cakes baked on board were served at tea time. Supper was a cold meal washed down with beer, rum or whiskey.

Still pushing against a headwind, they approached Belle Island at 8.45pm, the beam from its lighthouse visible off the port bow. The R100 was only the fourth airship to make it across the Atlantic Ocean, and there was plenty of fuel remaining as it headed down the Strait of Belle Isle on six engines, making good at 40knots (46mph/74km/h). At 6.25am on the Thursday morning they began calling the airfield at St Hubert as preparations were made for arrival. Eager to land before darkness, Scott increased speed to 70 knots (80mph/130km/h).

By 12.45 EST the airship was two miles (3km) off Father Point. Then the squall hit. Winds funnelling between the hills at the mouth of the Saguenay River met the cooler air spilling down from the mountains, and the resulting swirl of air rolled the airship sideways. The engines were slowed immediately and a party was sent to check the framework for any signs of damage. They discovered rents in the fabric on the lower and starboard fins; nothing too serious as long as it was immediately repaired to prevent further rips being caused in the airstream. Captain George Meager was a member of the inspection team, and moving to the port fin he discovered more serious damage. 'What a sight met my eyes: the fabric forming the underside of the fin just abaft the leading edge was literally in ribbons, and there was an enormous hole gaping in the underside of the fin, large enough to drive a double-decker bus through...'

A team of riggers set about the repairs, but it was two hours before the engines

The R100 approaches the mast at Saint Hubert's airfield, near Montreal.

could be opened up again to half speed, putting an end to any prospect of landing that day. Continuing to battle the headwinds, the airship reached Quebec late that afternoon as the sun disappeared behind banks of angry thunder clouds. At around 7.40pm they encountered storm conditions, abruptly thrusting the nose skywards one moment and then downwards the next. Strong air currents whisked R100 from 1,200ft (360m) up to 3,000ft (900m) in less than a minute. As it settled momentarily, another updraught grabbed the hull like an invisible hand and catapulted it to 5,000ft (1,500m). Throughout the airship, from the galley to the wireless room, every moveable object had been pitched forwards, and inevitably there was more damage to the fins.

The R100 flew over Montreal in darkness at a little after 1.00am on the morning of 1 August 1930, and as dawn was breaking it reached its destination, Saint-Hubert Airfield. Despite the earliness of the hour the airfield and all the surrounding roads were alight with the beams of car headlights as thousands of Canadians flocked to see this flying Goliath. The 3,870-mile (6,280km) journey from Cardington had been completed in 78hr 49min, and 1,505gal of fuel remained. The R100 and its crew stayed in Montreal for twelve days, during which time over a million people came to see it at the airfield. Countless more got the opportunity when it made a twenty-four-hour flight taking in Ottawa, Toronto and Niagara Falls.

The R100 departed on the homeward leg across the Atlantic on 13 August and reached Cardington after a record-breaking flight of just under 58hr. A huge crowd turned out to welcome the airship home, and as Major Scott and his crew emerged from the base of the mooring mast the Minister of Aviation, Lord Thompson, greeted them warmly. Thompson was an ambitious career politician with an eye on becoming Viceroy of India. He was adamant that the R101 would follow R100's example, and should be ready to fly to India and back in time for the Imperial Conference being held in London in October. Nevil Shute has suggested in *Slide Rule – Autobiography of*

Canadian aviation officials welcome the crew of the R100 at St Hubert Airport, Montreal, on 2 August 1930. Left to right: Sir Dennistoun Burney, designer of the R100; Colonel J.L. Ralston, Canadian Minister of National Defence; Major Scott, and Wing Commander R.B.D. Colmore.

The R100 over Montreal in August 1930.

an Engineer that the success of the R100's double Atlantic crossing contributed to the downfall of the R101 by creating a climate in which failure of the state-built airship to fly to India was no longer an option.

Imperial ambitions

After its return from the Canadian voyage the R100 was placed in the shed while work proceeded on R101. Both airships remained under evaluation, and a document uncovered at the National Archives by researchers from the Airship Heritage Trust has thrown new light on improvements being considered. This Air Ministry note to the Director of Airship Development, dated 28 August 1930, begins by acknowledging that neither airship had met the original requirements to ascertain the viability of a regular international service because both lacked sufficient lift. To make any progress it was necessary to 'Carry out improvements to R100 and R101 such as modifications to passenger accommodation in both ships, fit "dieselized" Condors to R100 and possibly fit a new bay to the ship'. Placing some of the passenger accommodation externally beneath the keel would have made room to

Standing at the base of the mooring mast, Lord Thompson congratulates Major Scott on the R100's safe arrival back at Cardington.

increase the volume of the gasbag immediately above it. Furthermore, if this approach was carried forward to the R102-class it might have been possible to fly with a larger number of passengers than originally envisioned, probably fifty.

Both the R100 and R101 were intended to become the workhorses of the England-to-Egypt route, for which an additional mast was erected at Malta. There would also be occasional flights to Canada and India to maintain interest in those countries. The

Major Scott poses for photographs at the window of R101's control car.

R102 and R103 were to be the first of the next generation of airships, scaled-up versions of the R101. The R102 was to have a volume of 8,300,000cu ft (235,000cu m) and the R103 would be 9,500,000cu ft (268,850cu m). The development programme also called for the lengthening of the existing sheds at Cardington and the

construction of a new mast and a shed capable of housing two ships side by side, as well as undertaking the improvements to R100 and R101. The completion of the first of the new ships was to be within three years. All of these airships would need trained crew, but as things stood in the autumn of 1930 there was already insufficient personnel to man both existing ships at the same time, let alone an extended fleet.

R101 – the end of the dream

With R100 in the shed, every effort was directed to getting the R101 ready for the flight to India. Lord Thompson was champing at the bit, as revealed in one of his frequent memos: 'I must insist on the programme for the Indian flight being adhered to, as I have made my plans accordingly'.

Still lacking sufficient lift despite the extension, the R101 remained woefully untested for such a challenging voyage. Only days before the scheduled departure the required forty-eight-hour trial was hastily replaced by a sixteen-hour flight in perfect

The R101, with the civil registration G-FAAW, drops water ballast beside the Cardington mast.

conditions. Yet despite the misgivings of Colemore and his people concerning the chafing of the gasbags, leaking gas valves and the condition of the outer cover, Lord Thompson got his way. At 6.30pm on the evening of 4 October 1930 R101 backed away from the mast and disappeared into the darkness. To many of the onlookers it appeared to be struggling to gain height, and the dropping of around four tons of water seemed to confirm that it was heavier than anticipated.

The flight schedule called for a stopover at Ismailia, Egypt, the next evening, then proceeding to Karachi the following morning. On board was a crew of forty-two plus twelve passengers, and as they settled down to supper the airship was already encountering increasing headwinds that caused it to roll gently, like a ship. Once over French soil course was set for Paris, and by the time of the watch change at 2.00am it is thought that both Major Scott and Flight Lieutenant Irwin had left the bridge to catch up on some sleep. Making little headway, possibly only 20mph (30km/h) ground speed, the R101 suddenly lunged down-

Loading luggage before the R101's departure for India, with Flying Officer Steff handing over some suitcases.

wards, righted itself and then dived again, striking a hillside near the small town of Beauvais. Within moments it was engulfed in flames as the hydrogen burned, killing forty-eight men in the inferno. Eight managed to clamber clear of the wreckage, but two of them died of their injuries over the following days. The cause of the crash

The crew of the R101 with the airship riding at the mast. Many of these men died in the accident which destroyed the airship on 4 October 1930. Four of the six survivors, all engineers, are clustered in the middle, with V. Savory fifth from the left, then Joe Binks, Arthur Bell and Alf Cook.

The dream of an Imperial airship service died with the crash of the R101 at Beauvais in northern France. This edition of the News Chronicle of 6 October 1930 states that eight men survived, but two died of their injuries shortly afterwards. The wreckage lay on the hillside well into 1931, when scrap contractors from Sheffield were sent to salvage what they could.

has remained a matter of conjecture, but the most likely scenario is that a failure in the outer cover near the bow initiated the sudden dives. The fact that the airship was heavy and fighting difficult weather conditions had compounded the situation.

The loss of the R101 was a spectacular demonstration of political aspirations and pressures stretching sensible caution beyond breaking point. In one fatal moment Britain had lost the cream of its airship men, and any real prospect for the continuation of its rigid airship programme. The R100 remained in its shed while the government debated its future. One option was to refurbish the cover and to continue with the projected R102, or perhaps to keep the R100 grounded until future plans had been finalized. Eventually the lack of funding at a time of economic depression persuaded the British government to abandon the Imperial Airship Scheme altogether, and work on dismantling the R100 began in November 1931. The framework of this fine airship was sold as scrap metal for just £427.

A Punch cartoon published shortly after the R101 was destroyed, showing the Spirit of France comforting Britannia. 'In grateful recognition of the spontaneous sympathy shown by France in the hour of England's sorrow.'

Relics of the Imperial Airship Scheme

Many reminders of the former British rigid airship programme are still to be found at Cardington, just a few miles to the southeast of Bedford. Most prominent are the two airship sheds where the R101 was built and R100 was housed. No.1 shed is in need of repair at the time of writing (2009), while No.2 is in far better shape and has been used in recent years by Warner Brothers for the filming of Gotham City in the *Batman* film series. The tall mooring mast is long gone, demolished during the Second World War for its metal. Beside the main road, the A600, the former red-brick office building for the Royal Airship Works still stands tall and proud, although it is awaiting refurbishment as part of a scheme to redevelop land to the western side of the hangars for housing. Across the street is Shortstown, a housing estate built for the workers at Short Brothers, and later the Royal Airship Works. The bigger houses on the crescent at the front were for the senior management, while the workers lived in the smaller ones.

Perhaps the most poignant relic of this historic location is the simple memorial in the village of Cardington itself, marking the grave of the forty-eight men who perished on the R101. It is said that a million people lined

The mass grave of the forty-eight men who died in the R101 is situated in the village cemetery at Cardington. Most of the bodies were far too burnt to be individually identified.

the streets to watch the funeral cortege make its way from Bedford station up the hill to the churchyard. Among those walking behind the coffins was Hugo Eckener, representing the Zeppelin Company. Today, the R101's flag, which miraculously survived the fire, hangs in the church of St Mary the Virgin, across the road from the memorial.

The only overseas shed completed, the one at Karachi, was demolished after the war and sold as scrap metal.

The LZ127 *Graf Zeppelin* visited Cardington in 1930, and seventy-eight years later Zeppelin NT07 004 departed from there on 28 August 2008 on the first international airship cruise since the 1930s, stopping at Calais and then at Brussels on its way to the Netherlands. Shortly afterwards it was shipped to the USA to begin operations in the San Francisco Bay area. Cardington still remains a centre for lighter-than-air activity in the UK with Hybrid Air Vehicles, the company behind the SkyCat, located on the site. Both the Zeppelin NT07 and the SkyCat are covered in more detail in Chapter 16.

The imposing office building of the former Royal Airship Works still stands, awaiting redevelopment of the land to the west of the sheds.

CHAPTER 10

The *Hindenburg*

~∾ ∾~

Doctor Eckener's plan had been to follow on from the success of the LZ127 *Graf Zeppelin* with the LZ128. The *Graf* was too slow and too small for a regular North Atlantic service; the new airship would have been an expanded version of the *Graf* containing around 5,474,000cu ft (155,000cu m) of hydrogen and offering accommodation for thirty to thirty-four passengers. This larger size had only been made possible by the erection of a new construction shed at Friedrichshafen with clear inner dimensions of 820ft (250m) long and 151ft (46m) high. At around the same size as the British airships R100 and R101, the bigger Zeppelin would have been more commercially viable than the *Graf* and fitted with more powerful engines better suited to the North Atlantic conditions. But the loss of the R101 in October 1930 persuaded Eckener to abandon the LZ128. He resolved that the next ship would fly with non-flammable helium and would be powered by safer diesel engines. The Americans, however, had a virtual monopoly on this gas, which is extracted from certain natural gas deposits, and the Helium Control Act of 1927 prohibited its export.

The LZ129

The LZ129 was to be a truly magnificent craft. More than three times the length of a Boeing 747, its streamlined hull was 804ft (245m) long, large enough to contain a volume of 7,062,900cu ft (199,880cu m), and it would establish an unsurpassed level of luxury in transatlantic air travel. Above all others, this airship represented the realization of Count Zeppelin's and Hugo Eckener's dreams for the future of transatlantic flight.

Following the well established conventions of Zeppelin construction, the LZ129 had a duralumin framework. There were fifteen main rings, the four biggest ones amidships being of 135ft (41.2m) diameter, almost twice the girth of the old *Graf*. Main ring spacing remained at 49.2ft (15m), with two intermediate frames in-between. An increase in the number of longitudinal girders to thirty-six main, plus intermediates, gave the hull a far smoother finish than on the previous Zeppelins. As with the *Graf* there was a central corridor axial girder, but instead of Blaugas, which was no longer needed for the diesel fuelled engines, all of

the sixteen cells would contain helium. Instead of the traditional goldbeater's skin they were made from cotton coated with a new gelatinous material. Initially it had been proposed that inner hydrogen-filled core cells would be installed within the cocoon of helium to improve lift and to act as a disposable gas in compensation for the weight of fuel burned, but this innovation was later abandoned. A new 1,200hp diesel engine was developed for the airship by Maybach. Later known as the Daimler-Benz DB602, it gave a significant saving in fuel weight. Each of the four engines was housed in an external engine car, and each one drove a four-bladed pusher propeller.

The crowning glory of the new ship was its spacious passenger accommodation, arranged on two decks within the hull. On the upper A deck were twenty-five double-berth cabins, located centrally with large public spaces to either side. Here the passengers would eat in the modernistic dining area, spend their time in the lounge chatting or gazing out through the large inward sloping windows, or perhaps write post-

cards or read from a small selection of books in the writing room. They could even enjoy a cigarette in the special smoking room down below on B deck (not a first, for airships such as the R101 had beaten

Applying the special coatings to the outer cover. Some experts have suggested that the flammable mixture used may have been the actual cause of the conflagration.

the only protuberance beneath the hull. There were two main control stations. The rudder-man's wheel was positioned at the front, and he steered by reference to a gyro compass. On the port side was the elevator-man's wheel, and in many respects his was the most crucial role, for he was responsible for responding to and anticipating the airship's vertical movements, judging the angle of the ship from an inclinometer and its rise and fall from a variometer. The elevator-man's skill at maintaining smooth and level flight was considered a fine art, and it is said that the crew could always tell who was on the wheel from the smoothness of the ride. Above and to either side of the elevator position were toggles for the release of ballast and for valving gas. On the far side of the control cabin was the engine room telegraph, which transmitted signals to the egg-shaped engine cars to indicate 'idle', 'slow', 'full' and 'flank speed'. At the back of the control cabin were the chart and navigation rooms, plus a small utility room at the rear. From here a ladder led up to the wireless cabin within the hull, along with the officers' quarters, which had small external windows that are just visible in photographs, forward of the control gondola.

Politics and helium

As work on the new airship progressed, Germany's strained political situation inevitably began to impose itself upon the Friedrichshafen works and on the Zeppelin Company. While Eckener remained a staunch and vocal opponent of the Nazi party, he could not hold back the tide as their grip on political power tightened. Work on the mighty airship had started in 1931, but by 1935 Germany was at the height of a depression and the LZ129's construction was only completed through a nine-million

Zeppelin to it). This consisted of a room in which the air was slightly pressurized to prevent the admission of any hydrogen, and passengers entered via a swivelling air-lock door. The single electric lighter was woefully inadequate for cigar or pipe smokers, and all passengers were scrutinized by the bar steward on their way out to ensure that no sources of ignition were taken into other areas of the ship. The remainder of B deck comprised the functional facilities such as the kitchen, toilets, crew and officers' mess areas. There was even a shower room, a first for any airship. The passengers had to sign up to use it specific times, although by all accounts the flow of water was little more than a trickle. The more Spartan crew quarters were arranged just behind the passenger area on the lower deck.

The airship's compact control cabin was

At the Zeppelin Museum at Friedrichshafen there is a full-scale replica section of the Hindenburg's passenger accommodation, including the lounge and promenade areas.

The Hindenburg's dining area on the port side of the airship. The senior officers would usually dine with their passengers.

The cosy smoking room was situated on Deck B, admission through the air-lock entrance being closely monitored by the bar steward.

Advice to transatlantic airship travellers

At the beginning, it is hard to realize you are on board a Zeppelin; the comfort and protection from the weather, the spaciousness, the elegance and neat equipment of the well-appointed cabins, the courtesy and deference of the ship's company who are only too ready to help, awake in you a new conception of pleasurable travel.

To inform their would-be travellers and put them at their ease, the DZR provided them with a little booklet entitled *Airship Flights Made Easy*. This covered every aspect of their forthcoming experience; an early example of what we would now term 'FAQs', from booking the flight to baggage allowances, suitable clothing and what to expect on the voyage itself. Here is what it says about clothing:

The Airship Voyages Made Easy *booklet with advice for Zeppelin passengers.*

You will find that you do not need any special dress, because life on board an airship is similar to staying in a large hotel or on board a passenger liner. Lady passengers are well aware that a dozen frocks or gowns will weigh scarcely more than one suit of clothes for a man. But the difference in climate at the port of departure and that of the port of arrival should not be forgotten, and, therefore, it is advisable at all times to take with you a light overcoat.

The modern central heating and ventilation system installed on board the airship renders the change of climate almost imperceptible. One hint to the men; a lot of time is spent looking out of the window at passing ships and other scenes of interest below. Many will find a comfortable cap an advantage. The wearing of a dress-suit or dinner-jacket is, of course, quite optional. Nevertheless, we advise that one dark suit should be carried in the personal baggage for convenient and suitable evening wear.

The booklet concludes with three important rules to be observed by all passengers:

1. Not to throw anything overboard, as by doing so you may cause damage to the airship's propellers or hull.
2. Not to carry matches, automatic lighters, or to smoke in any part of the airship, except in the smoking room.
3. Not to leave the passengers' quarters except by permission and accompanied by a member of the ship's company.

In fact, passengers were actively invited on accompanied tours of the rest of the airship, usually in groups of three, although the only way from the passenger areas into the interior was via the Chief Steward's office on B deck.

The Hindenburg *'s passengers were drawn to the large windows of the promenade decks, in this case the one adjoining the lounge area on the starboard side.*

brushes and other articles are displayed upon the toilet stand. At once, your cabin acquires a homely personal atmosphere. You listen for the roar of engines, or the fierce rush and vibration of the air, but apart from a distant quiet murmur, everything is tranquil and peaceful. You feel that nothing will disturb your sleep.

All who flew on the *Hindenburg* were astounded by the comfort and smoothness of the ride. There was no sense of motion through the air, and only a minimal degree of pitching or rolling. It became a common party piece to place a pen on its end to see how long it would stay upright. Airsickness was not an issue when flying on the airship. One passenger, Richard Cooke, published an account of a flight in his 1936 book *Passenger by Air*:

The saloons provide ample lounging space for the fifty passengers which the ship is designed to carry. The port saloon contains the dining accommodation, cleverly isolated by a modernistic metal railing, the starboard holds a music room and small

writing room. A feature of the former is a very handsome grand piano; in the corner of the latter is a small stand for the sale of souvenirs, postcards and stamps. Both saloons are decorated in an ultra-modern style, in keeping with the essentially modern nature of the whole ship; simple backgrounds, plain, but bright colours and metal furniture form the chief ingredients. Both notable and delightful are the extremely clever etchings in the panels on

ABOVE: *Designed as a haven of peace and quiet, the reading and writing room was decorated with murals depicting the history of the world's postage services.*

the walls, ranging from a large illuminated map of the world to small sketches of scenes in various parts of the two hemispheres lying on the scheduled routes of the airship.

Weighing only 400lb (180kg), the special Blüthner grand piano was constructed mostly of aluminium and was covered in yellow pigskin. Interior decoration throughout the passenger areas was under the direction of Professor Fritz August Breuhaus of Berlin, while the mural paintings were the work of Professor Otto Arpke. Exact replicas have been painstakingly recreated in the full-size reproduction of the *Hindenburg*'s lounge at the Zeppelin Museum in Friedrichshafen.

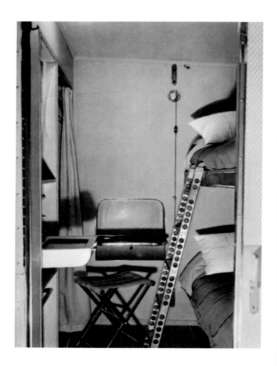

The interior cabins on the Hindenburg *were compact but functional enough. The table and sink could be folded away to allow more space.*

No one was ever bored when flying on a Zeppelin. As Eckener later wrote: 'To voyage in an airship high over the ocean is never monotonous, and is always delightful, because the picturesque scene of sea and cloud is always rapidly changing.'

As was to be expected, passengers enjoyed good food, served on fine china decorated with the DZR logo. The menu included Black Forest Brook Trout or Fattened Duckling, Bavarian Style, with Champagne Cabbage, or Venison Cutlets Beauval with Berny Potatoes, to be washed down with French or German vintage wines. The food was delivered to the dining room via a dumb waiter from the galley kitchen at the front of B deck. This was equipped with an aluminium electric stove with four electric rings, plus roasting and baking ovens for the making of fresh bread every day, a refrigerator and even an ice maker. It was usually out of bounds to the passengers unless they were on one of the special tours of the ship. Richard Cooke again: 'The huge kitchen, catering for a crew of forty as well as fifty passengers, is literally amazing; standing in the middle of this light and airy room, watching two cooks and a boy at work on the preparation of various meals, it becomes quite impossible to believe one is actually in the belly of an airship flying the Atlantic ocean.'

The 'Millionaires' flight'

While the *Hindenburg* was in the USA on its last northern Atlantic trip of the 1936 season, Eckener sought to cultivate the support of America's most influential industrialists by organizing a very special flight for an exclusive list of guests. These included individuals who might have some interest in the commercial future of the Zeppelin business, including representatives from various companies, plus post office and government officials from the USA and Germany, and reporters from the national press. The passenger list was a veritable who's who of the most influential names, representing a net worth of over a billion dollars, which gave rise to it being described as the 'Millionaires' flight'. These included the likes of Nelson Rockefeller, Winthrop Aldwich of the Chase Manhattan Bank and Paul Litchfield representing Goodyear-Zeppelin. On the morning of 9 October 1936 the *Hindenburg* departed from Lakehurst on its high-society outing. It flew over the George Washington Bridge, up the Hudson and reached the city of Boston by noon. Naturally for such a distinguished group the cuisine was outstanding, with Indian Swallow Nest Soup, fresh Black Forest Brook Trout, followed by Tenderloin Steak with Goose Liver Sauce, Carmen Salad and Iced Californian Melon served for lunch. The airship returned to Lakehurst at 5.22pm that afternoon with a party of well-fed and highly satisfied VIPs.

An American Airlines poster for the Hindenburg's service to Europe. The company provided air links to other destinations within the USA.

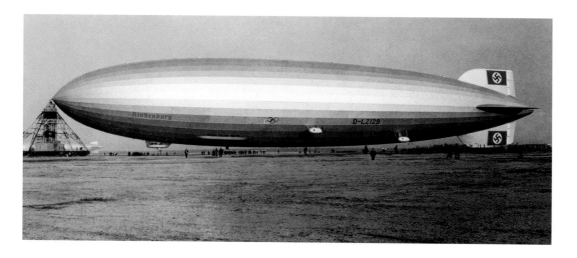

Registered D-LZ129, the Hindenburg is seen here at Lakehurst with the Olympic rings applied to mark the 1936 Games held in Berlin. The LZ126 Los Angeles is visible in the distance to the left.

A new age of airship travel

The *Hindenburg*'s first season had ended on an all-time high, and by this time work had already begun on a sister ship, the LZ130. The *Graf*, meanwhile, was kept busy servicing the South Atlantic route to Rio, and the brand new international operational base had been opened at Frankfurt-am-Main. Justifiably there was a tangible mood of optimism within the DZR, a sense that the golden age of the Zeppelin had dawned. It was with the following final words that Lehmann concluded his memoirs, completed at the end of 1936:

> The fight to conquer the air has ended in the defeat of the elemental forces of Nature. The era of inventions bows to the era of transportation. If airshipping has finally been established, it signifies more than the technical success – a triumph of perseverance... World traffic via airships has begun. By the expansion of weather-service on land and sea on the one hand, and the increased safety, comfort and practicality on the other, it will spread out over all the seas and continents. The world should be grateful to Germany as the trailblazer.

Despite the fine rhetoric there still remained the vexing situation of obtaining helium, with all supplies being blocked by the Americans. As it was evident that none was going to be forthcoming for the *Hindenburg*, its accommodation was altered over the winter lay-off of 1936–37

DEUTSCHE ZEPPELIN-REEDEREI

In 2 Days to North America!

SAILINGS and FARES
of the Atlantic Service

DEUTSCHE ZEPPELIN-REEDEREI
(GERMAN ZEPPELIN TRANSPORT COMPANY)

A DZR publicity brochure for the 1937 North Atlantic services.

to take account of the extra lift gained by continuing to use hydrogen, nine cabins being added to take the passenger capacity up to seventy-two.

CHAPTER II

Tragedy at Lakehurst

∽ ∽

On the evening of 3 May 1937 the LZ129 *Hindenburg* lifted off from Frankfurt on the inaugural North Atlantic crossing of the season. Eighteen flights to the USA were scheduled for that year. It was not the first flight of 1937, as the airship had already made a round trip to South America between 16 and 27 March, and was followed by the LZ127 *Graf Zeppelin* shortly afterwards. But it was for the North Atlantic route that the Hindenburg had been built, and on this flight to New York there were thirty-six passengers and a crew of sixty-one, including several trainees for the LZ130, which was under construction at Friedrichshafen. Hugo Eckener was not aboard.

Delayed arrival

The airship had been expected to reach Lakehurst, New Jersey, at 6.00am local time on the morning of 6 May, but persistent headwinds delayed its arrival and Captain Max Pruss, in command, radioed a new estimated time of arrival of 6.00pm in the evening. It was around 3.00pm when the inhabitants of New York City looked upwards, drawn by the drone of four power-

ful diesel engines as the *Hindenburg* passed overhead, making its way to the landing field at Lakehurst. It looked magnificent against the brooding clouds; the largest flying machine the world had ever known and the pride of the Zeppelin Company. Some may have recoiled with discomfort at the sight of the hated black swastikas on its tail, but few could have imagined that they were witnessing the last few hours of the great age of airship travel.

Captain Pruss managed to make up some

Several series of collectors' cards featuring the Zeppelin airships were produced for the German market.

IN 3 TAGEN NACH SÜD-AMERIKA!
DEUTSCHE ZEPPELIN-REEDEREI

DZR poster advertising flights on board the Hindenburg; '3 days to South America!'

time, and the *Hindenburg* appeared over the airfield at around 4.15pm, but no attempt was made to descend before the announced time, as arrangements for the ground crew and the various officials could not be changed at this late stage. This was unfortunate, as a weather front accompanied by rain and thunderstorms was approaching from the west. Pruss took the ship a few miles to the southeast to await the front's passage through the area. Nearer to the appointed time the airship approached Lakehurst again, coming in behind the weather front.

The Hindenburg *at the new international facility at Frankfurt, with the autobahn in the foreground. Both were potent symbols of a modern Germany.*

Commander Charles E. Rosendahl, the leading exponent of lighter-than-air flight within the US Navy, was in overall charge of the American ground crew, which consisted of ninety naval personnel plus 138 civilians recruited from the local area, who were paid a dollar a day. They were at readiness for 6.00pm, but with heavy rain and a thunderstorm still in the Lakehurst vicinity they would have to wait a little longer until the weather had cleared.

With the skies clearing a little, Rosendahl was finally able to radio Pruss, recommending that he make the landing, and at about 7.00pm the airship flew over the station at a height of 500 to 600ft (150 to 180m) so that he could take a closer look at the surface conditions. After circling once more the airship came back over the landing area, and after adjusting its trim and static conditions Pruss made a sharp turn, bringing it into wind and descending to about 200ft (60m) and 700ft (210m) from the mooring mast. His intention

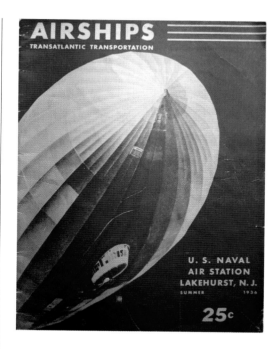

This souvenir publication was produced to mark the Hindenburg's *arrival at the US Naval Air Station at Lakehurst.*

was to make a 'high landing', an unusual practice but one which saved on ground-personnel costs. With the huge airship hanging almost motionless the engines were throttled back to 'idle astern'. At 7.21pm manila lines were dropped from hatches to the ground-handling party. On board LZ129 most of the thirty-six passengers were either standing by the large promenade windows on the starboard side of the lounge area, looking for their waiting families and friends on the ground, or packing their luggage ready to disembark. The Zeppelin's crew were at their various landing stations throughout the ship.

Rosendahl's men grabbed the ropes, connected them to the ground lines and began hauling them taut. In the wake of the thunderstorm the winds had fallen to light and variable, with just an occasional light gust pushing the airship slowly to one side. Four minutes after the landing ropes had been dropped the spectators on the ground noticed an inner glow and then a small burst of flame near the top of the airship, just forward of the upper fin. Rosendahl was one of them:

By 1935 Ernst Lehmann had become the director of the Zeppelin operations, and on the day of the Lakehurst disaster he was the most senior officer on board the Hindenburg, even though Max Pruss was in command. Lehmann suffered terrible injuries and died in hospital the following day.

It was a brilliant burst of flame resembling a flower opening rapidly into bloom. I knew at once that the ship was doomed, for nothing could prevent that flame spreading to the entire volume of hydrogen with which she was inflated. There was a muffled report and the flames spread rapidly through the after quarter of the ship.

Most eyewitness accounts report the sound as being more like the 'pop' of a gas stove than an explosion. Within seconds the rear of the ship began to drop towards the ground, overshadowed by a mushroom cloud of fire and smoke. In the control room the officers were not aware that anything was wrong at first, until they felt a shudder running through the ship and saw a yellowy-orange glow reflected in the windows. Flames poured through the passenger quarters as the floor tilted steeply, sending people and furniture tumbling towards the stern.

As the stern settled, the forward three-quarters of the ship, still having its buoyancy, pointed skyward at an angle of forty-five degrees. Through the axle corridor of the ship, in reality a huge vent extending along the central axis, the flame shot upward and forward as though it were going up a stack. Although the travel of the flame was actually progressive, it spread forward so rapidly and so quickly encompassing the entire length of the ship that to some it may have seemed almost instantaneous.

Pruss made a split-second decision to allow the airship to settle on the ground, ignoring his instincts to drop water ballast to slow its fall, but as the stern fell the nose reared up, spitting flame out through the nose via the central coaxial corridor before it began to settle. A few passengers jumped, and as the burning tangle of girders and fabric crashed to the ground others fought their way out. The ground-handling party fled in terror at first until their chief, Frederick J. Tobin, roared out: 'Navy men, stand fast! We've got to get those people out of there!'

A number of pressman were at Lakehurst that afternoon, as the inaugural flight of the season was still newsworthy. Among them was radio reporter Herb Morrison who was recording a commentary on the ship's arrival when the flames had erupted:

The ship is riding majestically towards us, riding as though it is mighty proud of the place it is playing in the world's aviation... There's smoke and there's flames now... crashing into the ground – not quite to the mooring mast. This is terrible. This is one of the worst catastrophes in the world... Oh, the humanity! And all the people are screaming around here... I don't believe it ... I can't even talk to people. Their friends are out there... I'm going to have to stop.

Survivors

Sixty-two of those on board the stricken airship survived, while thirteen passengers perished and twenty-two members of the crew died in the fire or from their injuries. One unfortunate member of the ground crew was also killed, bringing the total to thirty-six lives lost. There were many incredible escapes. Several men in the tail walked out unhurt, as the heat and flames spread upwards. One elderly woman

Consumed by fire, the Hindenburg falls to the ground at NAS Lakehurst, New Jersey, on 6 May 1937. Thirty-six people, including one member of the handling party, were killed in the disaster, but nearly two-thirds of those on board survived.

passenger walked down the airship's gangway in the usual way, while others clawed their way out with their bare hands. Joseph Spah, a professional acrobat, dropped 40ft (12m) from the promenade window, while thirteen-year-old cabin boy Werner Franz was saved when the contents of a water tank poured over him. Another passenger, Matilde Doehner, had thrown her two boys into the arms of the handling crew; the three of them survived, but her daughter and husband died of their burns. Mechanic Eugen Bentele was on duty in the forward port-side engine car when the first shock had run through the airship:

> My initial thought was that either a guy rope had snapped or one of the anchor points to the ship's framework had broken. When I looked down to see what was going on, I saw a fiery glow lighting up the landing field. There was no point jumping

Eugen Bentele, a mechanic aboard the Hindenburg, *was tending to the engine in the forward port car when disaster struck. With the airship's nose high in the air he was unable to jump at first, but as it settled to the ground he was thrown clear and managed to run to safety. He died in Friedrichshafen in 2003, aged ninety-four.*

at 60m [200ft] above the ground. The ship was already burning from its prow as far back as my engine car, and my main fear was that the heat would melt the aluminium framework so it would be unable to hold the huge weight of the engine car... It seemed like an eternity before the stern began to drop groundwards, slowly at first, then more quickly until the lower stabilizer fin struck the ground heavily... The ship's stern and other lower sections hit the ground and began to break up, so my engine car hit with less force.

Bentele was thrown over the top of the engine and out of the back of the car. After a moment of unconsciousness he came to, and then ran from the burning wreckage with only his hands and the back of his neck burnt. At a safe distance he turned around but could not see any other crew members. 'In front of me lay the wreck of our beautiful and proud ship.' Among the last to leave the burning wreck were the men in the control car. Ernst Lehmann had suffered serious burns and died a few days later, while Max Pruss survived, although he was also badly injured.

Causes and theories

The official inquiry, or rather inquiries, were fraught with political tensions. The Germans, reluctant to assist the Americans, whom they blamed anyway for not providing them with helium gas, held their own Gestapo investigation. But in an atmosphere of distrust and paranoia the truth, even if it was known, was unlikely to emerge. The investigators could not reach a conclusive explanation; only that the hydrogen had burned to such deadly effect. Evidence taken from the many witnesses, and from the film and photographic records, confirmed that the fire had started at the rear of the airship, just in front of the fin. Both the American and German investigators concluded that a mixture of hydrogen and air in the vicinity of gas cells 4 and 5 had been ignited by a spark. What caused this remains a matter of debate. Lightning, static electricity, the latent heat from a wire snapping under tension; or was it the flash of a saboteur's bomb?

Over the years various theories have come in and out of fashion. A bolt of lightning was out of the question; there were just too many eyewitnesses. Could the mooring lines being dropped to the wet earth have created a static charge? Hugo Eckener's suggestion that latent heat released by a snapped bracing wire, which had then sliced through the gasbags like a knife, was also considered plausible. The conspiracy theorists favour sabotage as the likely cause, especially as it makes for a far more dramatic ending in the vein of a Greek tragedy. Since that fateful day in May 1937 there have been countless books and films, all claiming to hold the truth despite the almost total lack of hard evidence. Others have even turned the conspiracy theory on its head by making the whole affair into a Nazi plot. But then that is conspiracy theories for you.

More recently one study pointed the finger of blame in an entirely new direction which, if correct, would mean that the mighty airship was brought down by some suspect paint. 'It was no hydrogen fire,' claims NASA scientist Addison Bain, an expert in all matters to do with hydrogen fuel. In 1997 evidence emerged from new studies which seemed to suggest that the special aluminium paint used to protect the outer cover of the airship was so combustible that under certain conditions it could be ignited by atmospheric electricity.

This stunning full-size replica of part of the Hindenburg's *passenger accommodation, displayed at the Zeppelin Museum, Friedrichshafen, conveys something of the scale of the airship. The ladder is an original one used in the airship's construction, while the nose cone of the LZ130 stands on the left.*

The coating of cellulose acetate butyrate (combined with nitrates and aluminium) was applied to waterproof and tighten the outer cotton fabric, as well as reflect the heat of the sun away from the gasbags. This particular concoction had never been tested on the previous Zeppelins; it was only used on the *Hindenburg*. Bain believes that the red-yellow colour of the flames, visible on some rare early colour footage of the disaster, indicates that the covering had caught fire before the hydrogen did. While this 'exonerating hydrogen' theory gained support among some in the airship world, it does not satisfy all of the experts and it is likely that the cause will remain a matter of contention. If Bain is right, the question is whether anyone within the Zeppelin company had realized the cause at the time. If they had, would they have admitted it and faced the wrath of the Nazi regime?

A view of Rio, looking forward from the rear engine car of the Graf Zeppelin. *By the time of the* Hindenburg *disaster the smaller* Graf *had been consigned to the South Atlantic route.*

The end of the dream

The *Hindenburg*'s destruction came as the finale to a spate of airship accidents around the world, and should have spelt curtains for the big airships. This was not the worst aviation disaster in history, not by a long shot, but the loss of the *Hindenburg* was one of the most public and spectacular disasters of the twentieth century. Through the combination of Herb Morrison's poignant commentary and the ghastly newsreel

The Graf Zeppelin *ended its days as a museum piece in one of the Frankfurt sheds.*

images, it has become indelibly etched on the public psyche for ever.

At the time of the accident the *Graf Zeppelin* was over the South Atlantic, returning to Germany from South America under the command of Hans von Schiller. He decided to keep the grim news of the *Hindenburg*'s destruction from his passengers until they had arrived safely at Friedrichshafen on 8 May. It is said that Schiller was prepared to take the airship out again on her next South America flight, scheduled for 11 May, but Eckener cancelled it. No paying passenger would fly in a hydrogen inflated airship ever again.

Over a nine-year period the *Graf* had performed impeccably, making 144 ocean crossings covering over 1,060,000 miles (1,706,000km) and carrying 13,110 passengers. It had been to the Arctic, flown around the world and become the only airship to cross both the Atlantic and the Pacific. But it was reaching the end of its useful life as it was, and was too small to fly with helium. On 19 July 1937 the *Graf* was flown up to Frankfurt, where it was strung up in the shed to be exhibited as a monumental curiosity for paying visitors while it awaited the final twist in its fate.

The US Navy's rigid airship programme

The USA's involvement with rigid airships seemed blighted from the start when the British-built R38/ZR2 broke in two during acceptance trials in 1921, resulting in the loss of forty-four American and British airmen. Five years later the indigenous ZR1 *Shenandoah*, basically a carbon copy of the First World War Zeppelin L49, was torn apart when it was struck by a line squall over Ohio in September 1925, killing fourteen of the crew. This left the US Navy with just the ZR3 *Los Angeles*, the former Zeppelin LZ126, which had arrived in the USA in October 1924 and enjoyed an exemplary flying career until it was decommissioned and dismantled in 1939.

In 1926 the US government authorized the construction of two identical rigid airships for the US Navy, and after a design competition the Goodyear-Zeppelin Corporation won the contract to build both the ZRS4 *Akron* and ZRS5 *Macon*. Designed by former Zeppelin engineer Dr Karl Arnstein, these had a volume of 6,500,000cu ft (184,000cu m) and at 785ft (239m) long were the largest helium-filled airships ever built; just 20ft (6m) shorter than the *Hindenburg*. With many innovative features, including swivelling propellers driven by internally mounted engines, plus a water-recovery system, their main function was to operate as airborne aircraft carriers. Both airships were capable of launching and retrieving up to four Curtiss Sparrowhawk fighters in flight via a trapeze system lowered from a hangar within their hulls.

Launched in 1931, the *Akron* was in service for two years before it was caught in severe weather off the coast of New England, causing its tail to strike the Atlantic waters. The ship broke up with the loss of seventy-three crew, and two more men died when their J-3 blimp crashed while searching for survivors.

The *Macon* was launched just two weeks after this accident, and had completed fifty flights when, in February 1935, it encountered bad weather off Point Sur, California. Violent wind-shears caused a structural failure at the point where the upper fin was attached, and as a result the rear gas cells were ruptured, causing the *Macon* to descend gently on to the surface, where all but two of the men were saved.

Despite this string of accidents, rigid airships remained on the US Navy's agenda for several years, although in the event the politicians had their way and no more American rigids were ever built.

The USA's flying aircraft carriers, the ZR4 Akron and ZR5 Macon, never did get to fly at the same time.

CHAPTER 12

The Nazi Airship

The destruction of the LZ129 *Hindenburg* at Lakehurst was not the end of the giant airships, but it did mark the beginning of the end. A year earlier, in June 1936, the keel had been laid down for a '*Schwesterschiff*' (sister-ship), the LZ130, which was to be an almost identical twin and has often been described as the second airship of the *Hindenburg* class. Arguably this was the finest airship ever built for the transatlantic service, but shockingly the LZ130 flew a total of only thirty times in its short career and never once across the 'Pond'.

To some extent Hugo Eckener had managed to protect the *Graf* and the *Hindenburg* from the influence of the Nazi Party, even though they had worn the swastika as the national symbol displayed by all German aircraft at that time. However, that was to all change as the National Socialist Party took far greater control of the final great Zeppelin, the LZ130.

The sister ship to the Hindenburg was the LZ130, which was christened Graf Zeppelin and is widely referred to as the Graf Zeppelin II.

THE LZ-130

This highly detailed cutaway of the LZ130 was published in Fortune magazine as part of a portfolio of craft flying the Atlantic in the future. Also represented were various aeroplanes, including Boeing's big flying boats.

The 'Schwesterschiff'

Construction work on LZ130 was well advanced by the time of the *Hindenburg* disaster, and the lessons learned were applied to the new airship. These included a revised formula for doping the cover, the application of graphite to the lacings on the cover to make them better conductors to reduce the risk of sparks, and greater monitoring of the electrical gradient between ground and airship before landing. But this was not enough for Eckener, who vowed that no passengers would ever be carried in a hydrogen-filled airship again. Following the conclusion of the official investigations he travelled to Washington to ask President Roosevelt personally to make helium available to the Zeppelin Company. In the immediate aftermath of the accident, public opinion in the USA, and that of Congress, had turned in favour of supplying helium, consciences having been pricked by the graphic images of the *Hindenburg*'s downfall. Roosevelt gave Eckener his word that helium would be forthcoming, and the Helium Control Act was amended accordingly. In June 1937 Eckener returned to Friedrichshafen in high spirits and work began on modifying the LZ130 to accommodate the new lifting gas. Because helium is slightly heavier than hydrogen it has a lifting capacity of only 93 per cent in comparison. Therefore considerable weight-savings would have to be made. Furthermore, helium is much more expensive than hydrogen, not to mention difficult to replenish. Therefore measures were needed to avoid unnecessary venting of the precious gas, primarily by the use of water recovery systems fitted to the engine exhausts to replace the weight of fuel being consumed.

Completion of the new airship was originally anticipated for September 1937, and DZR advertised that it would fly to South America in late October that year. The publicity photographs were taken, colour brochures printed and, despite the loss of the *Hindenburg*, the bookings still flooded in. Lifted by hydrogen, the LZ130 would have flown with seventy passengers, but the reduction in lift with helium meant the number of cabins was reduced. Originally the layout of the passenger decks had closely followed that of the *Hindenburg*, but now they were rebuilt for just forty passen-

A whimsical vision of Stuttgart's future Zeppelin port, with the Stuttgart-to-New York airship taking on passengers.

gers and rearranged on one main level. The area that had been the dining room on the port side was split into two rooms to provide a lounge and a smoking room, while the corresponding space on the starboard side was divided into a second lounge plus four

LEFT: This patriotic card depicting the LZ130 capitalizes on the legacy and popularity of Count Zeppelin, even though he had been dead since 1917.

new cabins. The dining room was relocated to run across the ship, with access from both lounges. There were twenty double cabins in total, of which thirteen had outside windows, an enormous improvement on the internal cabins of the *Hindenburg*. Photographs taken aboard the LZ130 reveal a series of modernistic, airy public spaces and well-appointed cabins.

To avoid venting helium in flight a system was developed to recover water from the engine exhausts and thus maintain the levels of disposable ballast for long-duration flights. This resulted in almost the only noticeable external difference to the *Hindenburg*, the four Daimler-Benz DB602 diesel engines driving tractor propellers at the front of the power cars, the rear ends of which were extended and tapered to accommodate the new equipment, which added six tons to the airship's weight. Weight savings had to be made elsewhere, and these included removal of two of the access ladders to the axial corridor running

LEFT: This patriotic card depicting the LZ130 capitalizes on the legacy and popularity of Count Zeppelin, even though he had been dead since 1917.

The modern interior of the LZ130, with the lounge area leading up to the wide dining room.

One of the passenger cabins with an outside view, although this photograph was clearly taken with the ship on the ground and quite possibly still inside the shed.

The LZ130's dining room, with a photograph of Nazi Air Minister Hermann Goering on the wall.

*The well-equipped
galley.*

through the airship's centreline, elimination of the manoeuvring valves, and improvements in the materials used on the gas cells. Almost eleven tons were saved, although this was reduced to between five and six tons once the added weight of the water recovery system was taken into account.

*An engineer tends to his
Daimler-Benz DB602
diesel inside one of the
noisy engine pods.*

Have airship, will travel

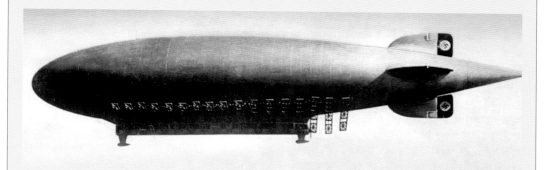

A scale model of the unfinished DLG or 'Speyer' airship of the mid-1930s.

This photograph from the mid-1930s shows a model of a little-known German project to build ten airships which were to be flown across the Atlantic to the USA. Sometimes referred to as the 'Speyer' airship project, it was the brainchild of businessman Otto Brinkman who set up the Deutsche Luftfahrzeug Gesellschaft (DLG; German Aircraft Corporation) at a former First World War aircraft factory in Speyer, a small town to the south of Mannheim. Initial announcements published in 1935 were for a non-rigid design with a volume of 706,200cu ft (20,000cu m), although this was later increased to 741,500cu ft (21,000cu m) when the first airship went into production. As his chief designer Brinkmann appointed Nikolaus Basenach, an experienced and respected engineer who had previously been associated with Major Gross in the building of a series of prototype Gross-Basenach airships for the Prussian Army Airship Battalion.

Information on the DLG airships is scarce, but Harold Dick, who was in Germany as Goodyear's representative at Zeppelin at the time, recounts several visit he made to DLG's works in 1936 in his book *Graf Zeppelin & Hindenburg*. On his first visit he found that work on a shed had started,

and he saw evidence of the 125ft (38m) gondola and the fins under construction, while the envelope was being made by the August Riedinger Balloon Factory at Augsburg. At 318.3ft (97m) long the streamlined envelope resembled that of the LZ120 *Bodensee*. Two cruciform girders would support the fins and power was to be supplied by three 365hp Junkers L5 engines, which were to be connected to propellers on outriggers. An interesting feature, clearly visible in this photograph of the DLG-1 model, is the 'nightsign' display system, which consisted of rows of letters formed by light bulbs mounted on seventeen racks running along either side of the envelope. (Goodyear was working on a similar system.) Once completed, the airships were to be ferried via Spain, the Azores and Bermuda to the USA. where they would be used as advertising airships.

Having spent some time with Brinkmann, Dick quickly surmised that, while this was a genuine project, its primary purpose was to get around the Nazi's block on the movement of money out of the country. Unfortunately, by the end of 1936 Brinkmann had probably been rumbled by the authorities, and with the company starved of funds the DLG-1 was never completed.

The struggle for helium

During the latter part of 1937 the arrange-ments for getting the helium to Germany were well in hand. It was to be shipped in high-pressure gas cylinders from the port of Galveston, Texas, then taken by barge up the Rhine for final transportation by rail to Friedrichshafen, or possibly by pipeline direct to Frankfurt, where storage facilities were prepared to hold up to 7,062,000cu ft (200,000cu m) of the gas. As Eckener saw it, everything was proceeding according to plan for the return of the transatlantic Zeppelin service. After all, Roosevelt had given his word, and despite delays the LZ130 should be flying by April 1938. Even the Nazi Air Minister, Hermann Goering, who had previously professed that he was 'never much for airships', had publicly vowed to re-establish the air link between

The nose cone of the LZ130 on display at the Zeppelin Museum, Friedrichshafen, is a rare remnant of the last great Zeppelin.

Germany and the USA. Such had been the groundswell of public support in favour of the Zeppelins that Goering had come to rec-ognize their importance as symbols of German pride and prestige. Much in the same way as with the LZ4's Echterdingen incident in 1908, the loss of the *Hindenburg* had resulted in an outpouring of popular support, and financial contributions began to flood in from the German people, to the extent that the government clamped down on the notion of a public fund lest it enhanced Eckener's national standing at the expense of the Nazi Party.

The climate of optimism for the future of the Zeppelins was at such a high level that in October 1937 it was decided that the next one, the LZ131, would be increased in size to carry a more economically viable eighty passengers by increasing the length of the hull by 59ft (18m). This was to be done by adding a 54ft (16.5m) bay and extending another. It would have been the biggest Zeppelin so far, with a volume of 7,994,750cu ft (2,262,500cu m) and an overall length of 863ft (263m). Another mark of confidence was the construction of a second shed at Frankfurt, at right angles to the '*Hindenburg*' shed (now occupied by the original *Graf Zeppelin*). This was completed by late 1938.

All these plans were thrown into disarray when it became clear that, once again, the Americans were not going to let Germany have any helium. Under the American polit-ical system the unanimous support of all members of the Munitions Control Board was required to approve its export, but the Secretary of the Interior, Harold L. Ickes, refused to vote in favour because he feared that Germany might use the airship for mili-tary purposes. When Wehrmacht troops marched into Austria on 12 March 1938 to enforce the *Anschluss* or 'union' with

Germany, it served to harden attitudes in the USA, and Ickes quickly gained both public and political support for his stance. By the time Eckener returned to the USA the following month to meet with Roosevelt again, the President was not willing, or able, to help.

This was a bitter blow to Eckener, who was left with a massive white elephant on his hands. It might have been the world's finest airship, but it could not be used for passenger flights. Berlin gave approval for the LZ130 to be inflated with hydrogen so that it could be used for propaganda and spying flights, and the airship was remodified accordingly. Eckener, meanwhile, clung on to the hope that a series of successful demonstration flights might just cause the American government to relent and release some helium. A non-stop return flight to the USA was a possibility, but the Reich Air Ministry refused permission for the airship to be flown outside of Germany's borders.

Graf Zeppelin II and the spy flights

On 14 April Hugo Eckener christened the LZ130 with a bottle of liquid air, naming it *Graf Zeppelin*. His words were a poignant indication of the sway that the Nazis now held over all aspects of life in Germany: 'I wish this ship to carry forth the honour of German technology to all the world as a symbol of unshakable German will, to gain for the German people the place in the world to which they are entitled. To the German people and their leader! Sieg Heil' It could have been written by Goebbels. It probably was.

Although the LZ130 is widely referred to as the *Graf Zeppelin II*, there was no confusion at the time, as the original LZ127 *Graf*

The *LZ130* Graf Zeppelin II *flying above the shed at Frankfurt.*

Zeppelin was ensconced within the shed at Frankfurt, serving out her days as a museum exhibit. Apart from Zeppelin Company officials, only Hermann Goering of the Reich Air Ministry was present at the naming ceremony. Immediately afterwards the LZ130 *Graf Zeppelin II* was walked from the shed and ascended for its ten-hour maiden flight, flying over Munich, Augsburg and Ulm before returning to Friedrichshafen. Six further trial flights were completed before it flew to the new operational base at Frankfurt at the end of October 1938 and was placed under the command of Captain Albert Sammt, who had recovered from the injuries he sustained at Lakehurst. The LZ130 then embarked on a series of propaganda flights

Members of the Luftwaffe's construction battalion inspect the empty shed in the spring of 1940.

on behalf of the Reich Ministry of Public Enlightenment and Propaganda, visiting various locations within Germany, plus a number of what can only be described as 'spy' flights, or, as we would refer to them nowadays, electronic espionage missions. Ironically these covert activities only vindicated US Secretary of the Interior Harold Ickes's predictions that the airship could be used for military purposes. A section of the passenger quarters was secretly converted to house special radio-detection equipment designed primarily to sniff out Britain's early-warning defences.

The most notable of these missions was also the LZ130's longest flight, lasting over forty-eight hours from 2–4 August 1939, when the airship flew northwards just off the east coast of Britain, up to the Shetland Isles and back. Unfortunately for the German technicians on board, the airship's metal framework tended to disperse the incoming signals, making it difficult for them to pinpoint the source, a problem exasperated by the Luftwaffe's refusal to halt its own transmissions. To obtain clearer readings an old observation 'cloud car' dating from the First World War was pressed into service. This was dangled several hundred feet beneath the airship, which just happened to experience an unexplained 'engine problem' near Aberdeen, causing it to drift inland and nearer to some interesting-looking antennae. Royal Air Force Supermarine Spitfires were immediately sent to intercept, and as soon as they were spotted by the Zeppelin's crew the observer in the cloud car was quickly hauled back into the airship. The British filed a diplomatic protest, accusing the Germans of spying, and on its return to Frankfurt the following afternoon an Allied delegation was waiting to inspect the airship. Captain Sammt was advised not to land until dusk, and to do so at the far side of the airfield, allowing time for the secret monitoring equipment to be spirited away out of sight.

The end of airships?

The LZ130's final assignment was to attend an air show at Essen-Mulheim Airfield in northern Germany on 20 August 1939. An estimated crowd of 250,000 people had turned out to see the airship. But despite the airship earning revenue from the Nazis for their propaganda and spying flights, as well as a share of the fees for attending such air displays plus money earned from carrying special mail for philatelists, it was not enough to save the last great Zeppelin. It returned to the shed at Frankfurt on 21 August, never to fly again. In the spring of 1940 Hermann Goering sent a Luftwaffe construction battalion to the airfield to break up the two *Graf Zeppelins* on the pretext that their metal could be put to better use. Then, on 6 May 1940, coincidentally the third anniversary of the *Hindenburg* disaster, the Frankfurt sheds were dynamited to make way for the Luftwaffe bombers using the airfield. Work on the frame of the LZ131 had already been

stopped during the previous summer, and *Flight* was not alone in raising the question of whether this was 'The end of Airships?', as reporter Miles Henslow commented:

> Were helium supplies available there is no doubt that airships could more than hold their own as a swift, safe and reliable means of transoceanic transport for at least twenty years. Now, however, every year wasted makes the case for airships more forlorn, and, while one cannot help feeling the greatest sympathy and admiration for the German airship men, Dr Eckener in particular, it looks as though their work has been in vain.

In truth, the notion that the airships might have ruled the skies for another twenty years was blown away by the accelerated rate of aeronautical progress in the white heat of the Second World War, especially with the development of the jet engine. While the destruction of the *Hindenburg* had been a fiery beacon marking the

Destroyed in seconds, one of the sheds at Frankfurt falls to make way for Goering's aircraft.

An engine car of the LZ127 Graf Zeppelin, at the Zeppelin Museum in Friedrichshafen.

airship's decline, it was the rise of the aeroplane that brought its inevitable demise. The clues had been evident for some time. When the *Hindenburg* made its first voyage to North America, back in 1936, the US Navy had produced a commemorative publication entitled *Airships Transatlantic Transportation* which was sold for twenty-five cents on behalf of the Navy Relief Society. Its pages were filled with images of the *Hindenburg*, including the construction and the stylish passenger decks, and a series of articles extolled the virtues of airship travel over the 'short-ranged' aeroplane.

As to the element of speed, the primary and preponderant reason for any form of air transport, the airship plugging along day and night non-stop, with no necessity for intermediate stops on even the longest of ocean stretches, on a 'great-circle' rather than a zigzag course, free to choose its path with regard to meteorological science and conditions because of its abundant cruising radius, even at its lower hourly speed... actually arrives at its destination in appreciably less elapsed time than does the short-ranged aeroplane.

But sandwiched between the fine rhetoric and photographs of the *Hindenburg* were advertisements for the airlines and aircraft manufacturers, snapping at the heels of the airships and hinting at the fast and efficient services of the future.

CHAPTER 13

The Blimp Bounces Back

When the Japanese paid their surprise visit to Pearl Harbor on 7 December 1941 the US Navy had only ten operational airships available to patrol its entire coastal waters. Ever since then the debate has continued over whether an airship presence at Pearl Harbor might have given sufficient warning to repel the attack more effectively. Either way, the damage was done, and when Hitler declared war on the USA just four days later it was clear that the US Navy was woefully underprepared for the ensuing U-boat onslaught against its surface vessels. It was not for a lack of good intentions, however, for Congress had approved the '10,000 Plane Program' in December 1941, with a provision for the procurement for forty-eight airships, a number increased to 200 in June 1942.

Blimps for the US Navy

Spurred into action, the US Government inaugurated a massive airship building programme alongside a large-scale training programme to provide the necessary personnel. In 1941 there had been only 100 airship pilots, but by 1944 this number

had increased to 1,500 pilots plus around 3,000 crewmen. The stalwart of this airborne fleet was the redoubtable K-class, although the US Navy also used others models such as the smaller L-ship, based on the Goodyear advertising airships, primarily as trainers. The K-ship, however, was purpose-built for its wartime duties.

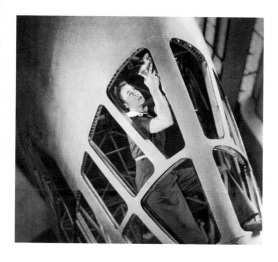

Just as in the UK, many women joined the workforce in the USA to aid the war effort and release the men for other duties.

The Resolute *was one of several former Goodyear advertising airships recruited by the US Navy in 1942. Redesignated L-4, it was assigned to coastal patrol duties, operating from Miami NAS. In this photograph the crew are being sworn into naval service. With the introduction of the bigger K-ships, the L-ships became the primary trainers. (US Navy)*

A US Navy blimp taking part in submarine hunting exercises. Their slow speed made them ideally suited for both visual sub-spotting and the operation of Magnetic Anomaly Detection (MAD) gear. (US Navy)

In addition to their roles as submarine hunters and shipping escorts, the US Navy blimps helped in the search for survivors of sunken vessels, dropping supplies and reporting their position to surface vessels. (US Navy)

Specifications for the K-ships varied slightly during the course of the war, in particular with an increase in the envelope size from a little over 400,000cu ft (11,320cu m) up to 425,000cu ft (12,028cu m) in the later models. Powered by two air-cooled engines, mostly 425hp Pratt & Whitney R-1340-AN-2 radials, they had a range of over 2,000 miles (3,200km), an endurance of around 38hr, a cruising speed of 58mph (93km/h) and a top speed of 78mph (125km/h). They were normally operated by an aircrew of ten; pilot in command, two copilots, a navigator, a rigger, an ordnance-man, two mechanics

The ghost ship L-8

On a quiet Sunday morning, 16 August 1942, the inhabitants of Daly City, California, were startled to see the distorted shape of a partly deflated US Navy blimp descending on to the town. It scraped across some power lines before coming to rest at an intersection. The local police and firemen dashed to the scene and, finding the gondola unoccupied, slashed the envelope in the mistaken assumption that the crew might be trapped inside. There was no one aboard.

The L-8 had taken off on a routine patrol flight from NAS Treasure Island in San Francisco Bay at 6.00am that morning with a crew of two, Lieutenant Ernest Dewitt Cody and Ensign Charles Adams. Both, incidentally, were survivors of the *Macon* crash in 1935. At 7.50am Cody reported to the airship base at Moffett Field that they had sighted an oil slick 5 miles (8km) off Farallon Island and were going to investigate. That was the last radio message ever received. Investigators examining the remains of L-8 found that the radio was operative, although the microphone was hanging outside the open door. The throttles were open and the parachutes and rubber raft were still in place.

Cody and Adams were never found, and over the years the mystery of the L-8 has attracted a number of theories. One eyewitness, the skipper of a local fishing boat, reported seeing the airship descending close to the surface of the water and dropping two smoke flares. The official conclusion

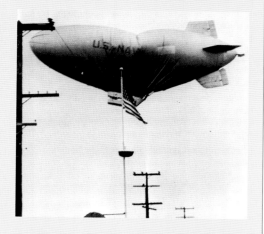

Losing pressure, the L-8 drifts over the power lines as it descends in a residential district of Daly City, California, in August 1942. No crew was found on board.

was that one of the men must have slipped out of the gondola when investigating the slick, and the other had fallen in an attempt to rescue his colleague, both men then being swept away by the strong currents. The lightened airship would then have ascended to the point where the automatic pressure relief valves opened and it gradually descended to drift back on to the shore.

The gondola of the 'ghost ship' L-8 was later refurbished by Goodyear to be used with its advertising airship N10A *America* in the 1960s. It is currently on display at the Museum of Naval Aviation at Pensacola, Florida.

ABOVE: A wartime advertisement for Foot Bros gears, as used on the Pratt & Whitney engines of the K-ships.

and two radiomen. The gondola, or car, as the Americans prefer to call it, was 40ft (12.2m) long, and with the engines mounted on outriggers it was a noisy and not particularly comfortable environment for the crew, especially on lengthy patrols. As their main duty was to detect enemy submarines they were equipped with ASG-type radar, sonar buoys and magnetic anomaly detection (MAD) equipment. Four Mk 47 depth charges were carried, two in a bomb bay and two externally, plus a Browning machine-gun in the forward part of the gondola.

Goodyear built 135 K-class airships, and they were deployed on antisubmarine warfare (ASW) duties in both the Atlantic and Pacific oceans and, later on during the war, in the Mediterranean as the so-called 'Africa Squadron'.

Blimpron 14

In May 1944 Blimp Squadron 14, or Blimpron 14 for short, was transferred from the US Naval Air Station (NAS) at Weeksville, Elizabeth City, in North Carolina to the newly created NAS at Port Lyautey in French Morocco, which was under the operational control of Commander Eighth Fleet. Its principal role was to provide MAD anti-submarine patrols across the narrow Straits of Gibraltar. Six K-ships were sent initially, and a further two followed in the spring of 1945. This was the first time that any non-rigid airships had made a transatlantic

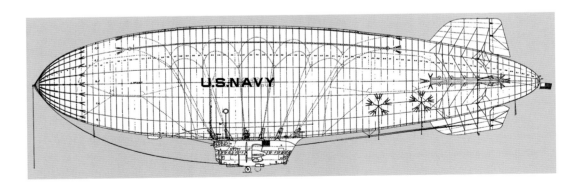

With a crew of nine or ten men, the K-class or K-ship became the lighter-than-air workhorse of the US Navy during the Second World War. A total of 135 of these 251ft 8in (77m)-long non-rigids were built.

enough for the British to relax the blackout conditions whenever two airships were operating at the same time, and on these occasions the running lights were shown.

Usually operating in pairs, the airships conducted their patrols at a height of 100ft (30m) as indicated on the radio altimeter. Each crew consisted of four pilots and six enlisted men, and to maintain alertness during the constant low flying the watches were changed every hour. The airships patrolled every night, except when extreme winds kept them at base, or when enemy aircraft were spotted approaching their positions. In the latter event they were required to switch off the lights and proceed at extreme low level, keeping close to the shore. It was anticipated that flying slowly would cause the enemy's attempts at radar tracking to become confused by surface vessels. The Blimpron 14 airships claimed several positive contacts with enemy submarines attempting to 'run the Straits'. And to test their efficiency the British submarine HMS *Voracious* made sixteen submerged runs and was located by the blimps on fourteen of them. French, Italian and British submarines were also used to conduct further exercises to refine the tactics and procedures.

With the coming of autumn Port Lyautey was frequently blanketed with morning fog which sometimes did not burn off until noon. This meant that the patrols were often extended by an extra four or five hours, the airships circling the airfield until the fog lifted. Accordingly a mobile mast was set up at Gibraltar to serve as an emergency landing site. On 18 July 1944 K-112 became the first US Navy airship to set down on the 'Rock', a difficult manoeuvre owing to the winds spilling around the high cliffs and the close proximity of the Spanish border. With fifty RAF men serving as a

With no sheds at Port Lyautey, repairs, such as this one to a lower vertical fin, had to be made with the airship riding the mast. Such work could only be done during periods of calm winds, and even then a mobile platform was essential. (US Navy)

makeshift handling party, K-112 was held down while the base commander climbed aboard for a spin, and after the second landing the airship was secured to the mast to be refuelled for its return flight to Port Lyautey. This diversionary destination was never used again.

Operating the airships from Port Lyautey was not without incident. On the night of 19 June K-130 struck the water while on patrol, resulting in the loss of the radar hat and radio altimeter antenna. In another incident, on 7 August, K-109 suffered a 16in (40cm) rip when it was struck by an aircraft's towing cable. Then, on 15 September, K-101 was almost wrecked when blasting operations carried out on the site hurled two heavy stones through its envelope. All

Blimp K-101 passes over the leaning tower of Pisa, northern Italy, in May 1945. (US Navy)

ballast and removable equipment was quickly unloaded to lighten the ship as the riggers swarmed over the wrinkled envelope to apply repair patches.

Mine-spotting duties

In September 1944 K-112 was assigned new duties in a joint operation with British Royal Navy vessels to spot and clear the countless mines that dotted the waters of the Mediterranean. The airship was sent to Cuers-Pierrefeu Aerodrome in southern France, a former airship base which had housed the two reparations Zeppelins LZ121 *Nordstern*, renamed *Méditerranée* by the French, and the former L72, which became the *Dixmude*. These rigid airships were long gone, but the twin 770ft (235m) sheds were still in reasonably good shape. The base had been liberated only eighteen days earlier, and it was being used by three Spitfire squadrons whose members were on hand to receive K-112 on the morning of 17 September. After the RAF moved out a US Army air/sea rescue unit moved in, and

shortly afterwards a group of German prisoners of war was assigned to the airfield to assist with ground handling and other duties. Although they were only 70 miles (110km) from the front, a greater threat to the airship came from the turbulent 'mistral' winds, which sometimes poured through the Rhone Valley at speeds of up to 110 mph (175km/h).

Aerial mine plotting began on 20 September, with the airship patrolling the coast and directing surface minesweepers to their location via radio and loudspeakers. In October, while K-112 underwent an engine change, K-109 took over its duties. Shortly afterwards K-109 was sent to Bizerte in Tunisia to carry out mine plotting. As the threat of the U-boats was declining, an advance base for Blimpron 14 was established at Cagliari, on the island of Sardinia, and the joint minesweeping operations with the British began from there in November.

The coming of peace

The last months of the war saw a couple of serious incidents involving Blimpron 14 airships. The first occurred in January 1945, when a French officer drove a Jeep too close to K-123 and snagged the emergency ripline, releasing the helium from the rapidly deflating envelope. Although a new envelope was ordered from Goodyear, this airship did not fly again until after the war's end. Then, in March, the masted K-109 was lifted on to its nose by a violent updraught, then crashed back to the ground, dragging the mooring mast across the airfield and damaging the engine so badly that it caught fire.

Two airships were now out of commission, and it was decided to send replacements from the USA even at this late stage of the war. On 28 April 1945 K-89 and K-114 departed from Weeksville on a very

One of the Blimpron 14 K-ships over Italy. (US Navy)

different course to the previous ships. They flew first to Bermuda, then on the main transatlantic leg from Bermuda to Lagens in the Azores, and finally from Lagens to Port Lyauty. The middle leg of the journey was the longest distance ever flown by a non-rigid airship, the 1,900 miles (3,060km) being covered in just over twenty-nine-and-a-half hours. It was on this leg that they were informed by radio of the presence of the German heavy cruiser *Prinz Eugen* dead ahead. Obviously the two airships were no match for the warship's heavy firepower or its Arado Ar 196 seaplanes, but they were already past the point of no return and, with inadequate fuel reserves for a major detour, had no choice but to plough on. Fortunately a thick fog bank covered the anticipated location of the warship, although it was later confirmed that the *Prinz Eugen* had never been anywhere near them.

Airships K-89 and K-114 touched down at Port Lyautey on 1 May 1945, just days before the end of hostilities in Europe, and the new arrivals joined the other airships of Blimpron 14 in their minespotting duties. They operated from bases at Cuers in

France; Cagliari in Sardinia; Rome, Pisa and Lido in Italy; Malta; and from the home base at Port Lyauty in French Morocco. During the last few months of the war only one airship had been retained to patrol the 'fence', and with the rapidly changing situation new roles emerged for the blimps, including convoy escort and air/sea search and rescue. In February 1945 K-112 escorted President Roosevelt's vessel to the Crimea Conference, and later that month K-109 escorted the President on the return voyage.

With the coming of peace the process of demobilization quickly gathered pace, and in November 1945 ZP-14 began preparations to go home. Sadly there was to be no return flight across the Atlantic, as the airships were crated for shipping. Most of the squadron's personnel departed from Marseilles on board the USS *Monticello* to arrive in New York on New Year's Day 1946, and Blimpron 14 was officially disestablished on 22 January. The eight K-ships had demonstrated that the Atlantic was not an insurmountable barrier to non-rigid airships, and during their time in north Africa and Europe they had saved many ships and countless lives. Unfortunately the lessons learnt in the US Navy's Second World War blimp were soon forgotten.

As a footnote to this episode of transatlantic operations, it is interesting to learn that the British had wanted a squadron of US Navy airships to be based in the southwest of England. Their role would have been to provide antisubmarine cover for the thousands of ships bringing essential supplies to Britain, but in the event peace came in Europe before the four K-ships were ready to be despatched.

CHAPTER 14

Dreams of Giants

∽ ∽

If the fiery destruction of the *Hindenburg* had sounded the death knell of the rigid airships, then the Second World War had left them well and truly buried. Such had been the headlong rush and scope of aeronautical development during the war, especially with bigger long-range aircraft and jet propulsion, that the passenger airship looked like a lumbering dinosaur by comparison.

In Germany the LZ127 *Graf Zeppelin* and LZ130 *Graf Zeppelin II* had been scrapped, the two sheds at Frankfurt dynamited by the Luftwaffe and the facilities at Friedrichshafen all but wiped out by the RAF because they had been used for the production of aircraft and V2 rocket parts during the war. Surveying the devastation, Hugo Eckener gave a brutally frank assessment of the prospects for his beloved transatlantic Zeppelins:

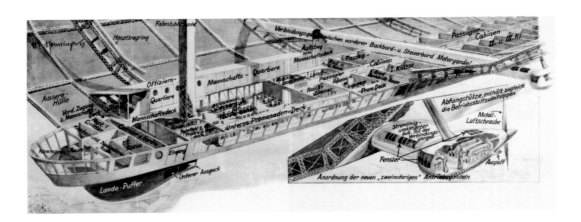

In 1929 the German magazine Die Woche *published this conceptualisation of a luxury passenger Zeppelin of the future.*

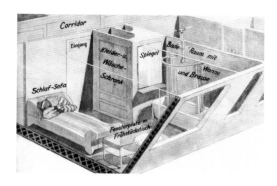

Die Woche's *1929 visualisation of one of the spacious passenger cabins, which includes a proper bed and even a shower.*

A tremendous war which, as always in life-and-death struggles between great peoples, had forced the rapid technical development of the weapons of victory without regard to cost and effort, had furthermore forced an enormous development and improvement in aeroplane performance, so that planes were now in a position to carry on a transoceanic service. The airship's monopoly was broken. And, since the aeroplane is much faster and can fly a given distance in half the time or less than is needed by an airship, the role of this aerial vehicle in commerce seems to have ended after a brief period of glory, just as it had been developed to the point of acceptance, for speed and time-saving are trump cards in today's hurried age... What does the airship have to offer now to the businessman or statesman in a hurry to cross the Atlantic Ocean?

'If you want to travel comfortably, take an airship'

Others were not so ready to relinquish the dream, and Max Pruss, the last captain of the *Hindenburg*, worked tirelessly to recruit financial backers to the airship's cause with his slogan 'If you want to travel quickly, take an aeroplane. If you want to travel comfortably take an airship'. Pruss's plan was to revisit the designs for the LZ131 to create a helium-filled airship powered by four 1,800hp diesel engines and capable of carrying 100 passengers at speeds of up to 100mph (160km/h). A bigger version, proposed in 1957, was for a 10,500,000cu ft (297,000cu m) airship which could transport either 70 tons of cargo or 200 passengers. Unfortunately Pruss's efforts were in vain.

In the USA it was a different story. The Americans still had their airship facilities, and in theory the rigid airship remained on the US Navy's agenda. During the war the Goodyear Aircraft Corporation, led by

'Air-Age Dirigibles' featured on the cover of Popular Science *in May 1945.*

Carl Arnstein, had produced a design for the 3,000,000cu ft (84,900cu m) ZRN training ship in anticipation of building a 10,000,000cu ft (283,000cu m) cargo ship by as early as 1947. But the Secretary of the Navy had not taken up the proposal, and by January 1947 any mention of a rigid airship programme had been quietly dropped from the Navy's policy statements. If Goodyear was to have any hope of keeping the airship dream alive, it had better start looking for backing elsewhere. In 1947 the company received some encouragement in the findings of the US Air Coordinating Committee (ACC), which had been considering the commercial future for lighter-than-air. Its membership consisted of representatives from the War Department, Navy Department, Civil Aeronautics Administration, Civil Aeronautics Board and the National Advisory Committee for Aeronautics. After two years of research they made a recommendation in favour of a rigid airship development programme.

Goodyear then began a high-profile promotional campaign in order to put its case to the American people and the politicians. Primarily this was through a series of advertisements in popular magazines, interviews in the press and the publication of a study of the characteristics of a large ocean-crossing airship. Written by Paul Litchfield and Hugh Allen of Goodyear, and published in 1948, *Why has America no rigid airships?* extolled the USA's need for the best possible transportation system, not only for the transatlantic route but also worldwide, to include both civil and military applications. At the heart of the proposals was an airship bigger than the world had ever seen before, 50 per cent larger than the *Hindenburg*, a commercial airliner with a helium capacity of 10,000,000cu ft (283,000cu m). The airship would be 950ft (290m) long, 142ft (43m) in diameter, would have an operating range of 6,000 miles (9,650km) with 35 per cent fuel reserve, and could carry a 90-ton payload over 2,500 miles (4,025km). Its maximum range would be in excess of 11,000 miles (17,700km). It would have a crew of forty, with perhaps twenty stewards, cooks and attendants, proportionate to the number of passengers on board.

Initially, Goodyear's advertisements featured the giant airships as flying aircraft carriers bristling with guns, but as the interest in the military application of airships waned the company turned increasingly to the postwar commercial potential. Many advertisements depicted the proposed rigid as a conventional design, the only obvious

'Cruise the world in a flying hotel' suggested a series of Goodyear advertisements in the 1940s. The emphasis was always on the quality of the well-appointed accommodation and not so much on the speed.

Goodyear's vision of the 950ft (290m)-long rigid airship of tomorrow.

new features being forward mounted elevators and an extended cabin under the hull, running for half its length. Passenger accommodation would be provided in one of three possible styles; a deluxe version with staterooms for 112 passengers, a Pullman layout for 232, or a low-fare model with reclining chairs for 288. 'Cruise the world in a flying hotel' they proclaimed: 'Want luxury... quiet... smooth sailing on you round-the-world cruise? Want a well-appointed stateroom, exquisite saloons, delightful meals and entertainment? You can have them all, one of these days – in great, safe, block-long airships!"

The key phrase in the sales pitch was 'one of these days'. But the well-groomed passengers seemed oblivious to the fantasy in which they were participating, and instead they admired the view from the panoramic windows, sipped their cocktails and chatted as they listened to a piano.

We're talking about the greatest airship ever conceived. It will carry more than a hundred passengers in uncrowded comfort. It will take you for long cruises to many lands – on your two-week vacation – bring you home as relaxed and refreshed as you'd be from a stay in the finest resort hotel. Airships are the quietest, most vibration-less means of transportation known. Airships offer nearly every advantage of an ocean liner – with three to four times the speed...

Newly strengthened light-metals and vastly improved fabrication – plus America's own non-flammable helium

The captain of the ship exudes experience and confidence while the younger officer looks on admiringly in this image from a Goodyear advert.

No airship lounge of the future would be complete without a piano to entertain the passengers on the forty-hour crossing.

gas – plus Goodyear's thirty-five years of experience in lighter-than-air craft – bring huge rigid airships well within the realm of definite post-war possibilities. But not as competitors of lightning-fast transport planes – nor any other form of land, sea and air transportation. No, airships will hold a well-defined niche that's all their own – luxurious, hotel-like comfort in long sustained flight.

Of course, nobody looking at the adverts could actually book a flight in one of these airborne hotels; it was all about selling the concept. But getting the public to believe in airships once again was always going to be difficult. The copywriters were attempting to damn the 'lightning-fast' aircraft with faint praise while emphasizing the airship's unique niche in the transportation market. In a final note smacking ever so slightly of desperation they inserted a familiar addi-

tional dimension to the argument. 'More important – airships can add another great arm to America's air-power. Militarily, as flying aircraft carriers and coast patrol ships. Commercially, as merchants and world travellers. Airships can be mighty ambassadors of peace and good-will – to every nation of the globe – and airships can help keep America first in the air!'

In a similar vein, further advertisements conjured up images of three-day flights to the Orient or transatlantic crossings in around forty hours, packed with 'entertainment, delightful cuisine' and 'luxurious relaxation'. One pictured the airship alongside its rivals – the passenger aircraft and ship – with Eddie Rickenbacker, the celebrated First World War fighter ace and president and general manager of Eastern Air Lines, providing his endorsement:

America needs all three... We need big, 350-mile-an-hour stratosphere planes – we need a great merchant marine and fleets of passenger steamers – and we need giant

The bright and modern interior of a cabin on Goodyear's futuristic 'London Express'.

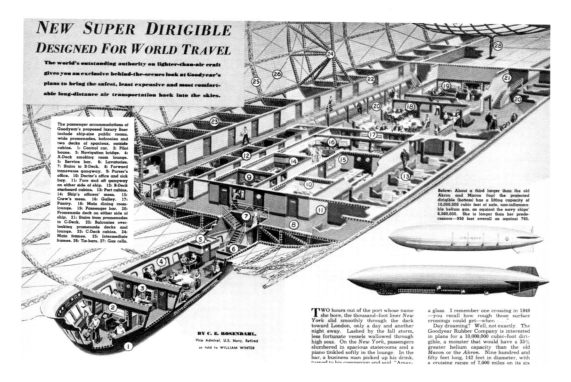

Another take on Goodyear's proposals for a 'New Super Dirigible – Designed for World Travel', published in Mechanix Illustrated, May 1948.

airships to round out the wide difference between the speedy plane and slow-going steamer. Each of the three has its own advantages – and each augments the other.

Support also came from Vice-Admiral Charles Rosendahl, now retired. In a 1948 interview he reiterated the arguments in favour of the airship.

When could we build such dirigibles? The answer is, today, now. In the United States we have the cream of the world's lighter-than-air scientists and technicians, many of them brought over before the war from Germany. The great dirigible dock at Akron, Ohio, still stands, as do the immense hangars at Lakehurst, New Jersey, and

Sunnyvale, California... To set up an adequate pioneering airship programme would cost this country less than it spends on a single airport such as LaGuardia Field. By building and operating dirigibles continuously, rather than as 'freaks', great progress could be made in this promising field. Why not faster dirigibles? As for bigger ones, engineers say a 15,000,000 cubic-footer (424,500cu m) is feasible, provided a big enough dock is available. How about jet propulsion at the stern, with engines and propellers mounted in internal tunnels or air passages? Or metal stitched-skin construction? These are but a few possibilities ... With this country's monopoly on helium, only America is able to build safe commercial dirigibles. These would be the biggest

The 'Atoms for Peace Dirigible' concept of the 1950s.

and safest profit-making aircraft, either heavier or lighter-than-air, that the world has yet seen.

It was going to take a lot of money if this dream was to become reality, and as Litchfield and Allen readily admitted, 'It would be impossible to find private capital to finance the project alone, carry it on to where it would be self-sustaining'. Establishing commercial airship lines would require the active interest of several branches of the US government, and none was forthcoming.

Nuclear powered giants

By the early 1950s the prospect of a new generation of large airships was given a fresh impetus. In 1946 the United States Air Force initiated the Nuclear Energy for the Propulsion of Aircraft (NEPA) project. The idea was that nuclear-powered bombers could remain airborne for extended periods, staying within striking distance of strategic targets. The programme had been intended to develop and test the nuclear-powered Convair X-6 aircraft. A reactor was test-flown on board a modified Convair B-36

to study shielding requirements and determine whether a nuclear powered aircraft was feasible, but although the reactor was 'hot' it did not power the aircraft. The programme was cancelled in 1961.

What had seemed like a good idea for heavier-than-air was embraced by the lighter-than-air community as an even better one for airships, as it might solve the problem of fuel displacement on long flights. The notion of atomic airships soon began to appear in the popular press. Taking its cue from President Eisenhower's suggestion that a nuclear-powered surface ship be built as a floating demonstration of the peaceful use of fission technology, *Mechanix Illustrated* proposed an 'Atoms-for-Peace' dirigible. A series of drawings by Frank Tinsley showed an elaborate design for a rigid airship 1,000ft (305m) long, complete with radar dome in the bow, a large four-bladed propeller at the stern and a series of retractable pontoons to enable it to land in the harbours of the countries it visited; although why it needed all this paraphernalia to land on water was never explained. If that was not enough, and to prove that some ideas never go away, it was also suggested that a helicopter landing pad and internal hangar deck might be installed atop the hull's centre section, along with promenade deck, nightclub and bar. There was even scope to accommodate an exhibition hall within the keel, to display 'examples of peaceful atomic age technology', which could be detached and lowered to the ground at various stops. According to *Mechanix Illustrated*, Arnstein and Eckener had personally commented on the design, saying that Goodyear's engineers considered it an intriguing new approach, and that the principle of taking the atom abroad via the airship was 'possible of technical accomplishment'.

A nuclear airship also featured in *The Zeppelin in the Atomic Age*, published by the University of Illinois in 1957, in which the author, Edwin Kirschner, reiterated the Goodyear and Atoms-for-Peace concepts and foresaw the potential for an initial fleet of four to six rigid airships operating between major cities, in particular across the North and South Atlantic: 'The longer the over-water route, the greater the advantage the airship has over the slower steamship and the more expensive aeroplane'.

Kirscher's book was intended to promote the broader case for a new generation of airships, and it offered little in the way of design specifics. During the 1960s Professor Morse of Boston University's Department of Aerospace Engineering took a far more considered approach to nuclear propulsion, and his team produced a design for a large 12,500,000cu ft (354,000cu m) rigid airship to be used for moving cargo or up to 400 passengers. The power plant would drive three 4,000hp gas-turbines turning two contrarotating propellers and a pair of 1,000hp turbofans designed to reduce drag. As was becoming almost customary, the control bridge was situated within the bow of the ship and the 400 passengers would enjoy 'hotel'-quality accommodation.

> Passengers would be housed in 42,000sq ft (3,900sq m) of deck space in staterooms like the best hotel rooms with fantastic views. There would be restaurants, shops, a theatre, cinema and games rooms in the lower compartment. On the top of the airship, reached by a lift, there would be a 'sky room' roofed with a dome of Plexiglas more than a hundred feet across. From here the passengers could enjoy the luxury of dining and dancing among the clouds.

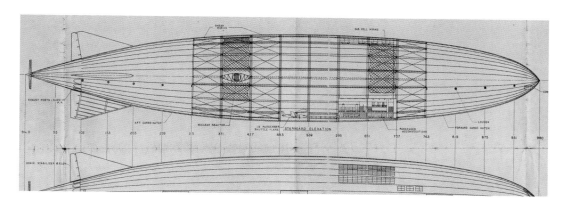

A blueprint for the atomic-powered rigid airship devised by Professor Francis Morse and his team at Boston University.

Not to be outdone, Austrian engineer Erich von Veress proposed an even bigger ship, the ALV-1, which would have a volume of 14,000,000cu ft (396,000cu m) and accommodate 500 passengers. Guy Hartcup, author of *The Achievement of the Airship*, actually suggests that neither of these designs was big enough to carry an adequate payload in addition to their means of propulsion. But such was the appeal of this Utopian vision of atomic airships with almost unlimited flying range that they were cropping up everywhere. The Soviet news agency Novosti issued images of what appeared to be a virtual clone of the Morse concept, but so packed with accommodation decks, helicopter and aircraft hangars that there was precious little space left for any lifting gas.

Throughout this period Goodyear continued to produce a number of airship designs for a wide range of applications, and with the rigid increasingly out of favour by 1959 its proposal for an atomic powered airship featured a large non-rigid for coastal patrol duties.

While nuclear energy had been heralded as the clean and efficient power source for the future in the 1950s, in more recent times, especially in the post-Chernobyl era, there is less expectation of nuclear powered aircraft or airships becoming acceptable. In 1984 Professor Morse wrote to me, expressing his revised views on the acceptability of nuclear power for airships: 'I agree that the concept is not currently viable, owing both to cost and safety considerations. But even before reaching this conclusion, I opted for a preferred system for long-range airships, one using cryogenic fuel (LH_2) which offers greater payloads than its nuclear counterpart.'

Having said that, some experts, including Professor Ian Poll of Cranfield University, have called for further research into developing nuclear propulsion as a viable alternative to fossil fuels. Admittedly there are advantages, such as being able to fly to the other side of the world without stopping and, significantly, the benefits of zero carbon omissions. However, the question remains whether nuclear radiation could be contained safely, even in a crash. The biggest problem facing the atomic lobby is

Passenger accommodation on the Morse airship included this 'Sky Room' within the top of the hull.

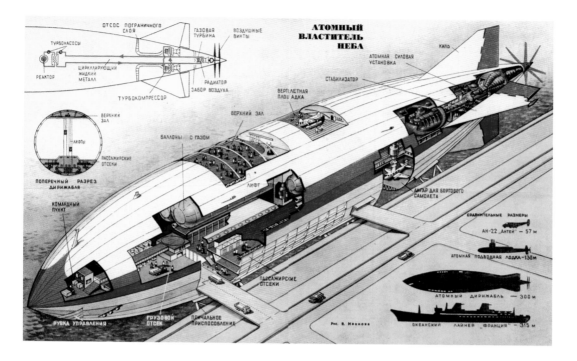

This imaginative diagram of a Russian proposal for an atomic-powered airship is so
packed with accommodation, helicopter pads and other paraphernalia that precious
little space remains for any lifting gas.

going to be the public's auto-response to the
nuclear issue. As Professor Poll told *The
Times* when interviewed on the future of the
aviation industry in 2008: 'We need a design
which is not kerosene-powered, and I think
nuclear-powered aeroplanes are the
answer beyond 2050. The idea was proven
fifty years ago, but I accept it would take
about thirty years to persuade the public of
the need to fly on them.'

Possibly of more immediate significance
in the future development of large airships
is the application of nuclear power stations
to extract hydrogen from sea water, either
as a lifting gas for future airships or even as
a fuel. But that is another issue entirely.

Returning to the real world of airship
operations in the 1950s, the truth is that,
while the dreamers fantasized over aerial
leviathans that were never to be built, it was
left to the lowly blimp to keep the lighter-
than-air dream alive.

CHAPTER 15

Extending the Envelope

Despite the US Navy's lighter-than-air organization emerging from the Second World War with an impeccable service record, the lowly blimps had not been directly credited with a single submarine kill, and their achievements were largely ignored in the postwar clamour for glory and a share of the limited defence budgets. The immediate aftermath of the war saw a sudden and unwanted surfeit of airships, and many were scrapped, mothballed or sold off to private ventures to serve as flying billboards advertising a diverse range of products and services. By 1946 the US Navy's

The record-breaking US Navy blimp Snow Bird.

lighter-than-air wings had been clipped to only two airship squadrons, ZP-12 and ZP-31, subsequently redesignated ZP-2 and ZP-1 respectively. Their mission require-ments were much in line with the wartime activities, including training, search and rescue, ferry and test flights, observation and photography and, of course, ASW. New units were formed, ZP-3 in 1950 and ZP-4 in 1951, to undertake development work and to evaluate systems that might be adaptable to airship operations.

Extended endurance

A small number of the K-ships that had been retained were upgraded with improved radar equipment, and bucking the decline there were the new larger N-ships. Known among the navy fliers as 'Nan ships', these had been in development by Goodyear from the late 1940s to fulfil the increasingly important AEW role as part of the USA's Cold War protective network. Unlike most conventional aircraft, airships can carry very large antennae and they make ideal platforms, being stable, vibration-free and able to stay on station for extended periods.

The initial version of the N-class was the ZPN-1, a follow-on from the wartime M-class airships, with a greater envelope capacity of 875,000cu ft (24,777cu m). In 1954 four improved N-class airships were delivered as ZPG-2s, although the Navy's designations varied with different versions, such as the ZPG-2W, which was modified for AEW missions. The envelope capacity for the larger ZPG-3W was 1,000,000cu ft (283,168cu m), roughly four to five times bigger than the advertising airships seen nowadays.

Determined to prove the value of its airships in a rapidly changing defence environment, the US Navy set about a series of extended-duration flights. Unlike the K-ship, the N-class had its twin engines mounted within the spacious gondola, enabling in-flight servicing during a mission, and refuelling at sea from surface vessels greatly extended tactical range. In May 1954 a ZPG-2 from NAS Lakehurst successfully completed an endurance flight of over 200 hours, travelling via Nova Scotia initially, then east to Bermuda and south to the Caribbean and the Gulf of Mexico and

finally landing at NAS Key West. The Chief of Naval Operations ordered a series of further tests to evaluate the airship's all-weather performance on long-duration patrols, and in 1957 another N-class airship maintained a continuous ten-day patrol 200 miles (320km) off the coast of New Jersey, experiencing some of the worst weather seen in decades, with freezing rain, icing, snow, fog and strong surface winds.

The next test was to be a headline-grabbing bid to beat the *Graf Zeppelin*'s 1929 record for the longest non-stop flight of 6,980 miles (11,233km). The plan was to fly a ZPG-2 named *Snow Bird* across the Atlantic and back again without refuelling or stopping. There had not been a trans-atlantic crossing by airship since the last pair of K-ships for Blimpron 14 were sent to Africa 12 years earlier, and no airship had ever achieved a non-stop double crossing.

The flight of *Snow Bird*

On the evening of 4 March 1957 the ZPG-2 *Snow Bird* was towed out of the shed at NAS South Weymouth in Massachusetts, ready to

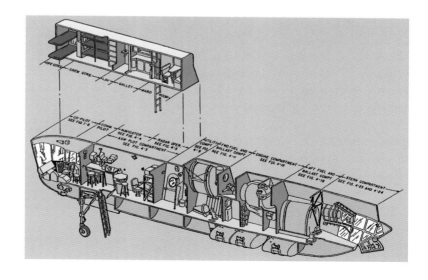

A diagram of the two-tier ZPG-2 gondola.

Airmen load food and other supplies on to the Snow Bird *in preparation for its long two-way flight.*

begin its historic mission. The late hour had been chosen to take advantage of the colder, denser air. As the ground crew completed the final preparations a smattering of light snow was falling on the airfield, but not enough to adhere to the envelope or cause concern. At 6.32pm Commander Jack Hunt gave the signal and the massive blimp lifted slowly, rising into the darkness and swaying slightly in the strong crosswinds.

On board *Snow Bird* the crew discovered that one of the bunk beds had been broken on take-off by the weight of the equipment stored on it, but this was the only minor problem. For several days the two-tier gondola had been stacked with supplies, including 1,017lb (460kg) of food, enough for fourteen men to last for at least twelve days. Some of the crewmen were amazed by the sheer quantity of frozen food carried aboard. In the late 1950s frozen food was still something of a novelty, and it was regarded with a degree of suspicion by many, who considered it overpriced and tasteless. Cases of canned supplies such as ham, turkey, milk and fruit were also

loaded. The dietary needs of the men had been an important part of the planning for the long journey, and the food's variety, weight and shelf-life were all taken into account. To ensure that the supplies lasted for the entire trip, a rationing system was devised, with each day's rations packed in separate bags. Water in particular was limited, especially when it came to washing or shaving, and the crew had been advised to bring an electric razor and deodorant. Every single item taken aboard the airship was carefully weighed. Each man was issued with a lightweight jacket and a base-ball-style hat, and before departure the kit bags containing dress uniforms and other clothing were taken to a Lockheed Super Constellation aircraft which would precede them on the various legs of the long flight. Its most important cargo was a portable mooring mast in case the airship had to make an unexpected descent.

The crew of *Snow Bird* comprised fourteen men. Commander Jack Hunt, the pilot in command, was supported by two copilots, Commander Ronald Hoel, the senior officer on the flight, and Lieutenant Commander

Lieutenant Stanley Dunton examines maps at the navigation station.

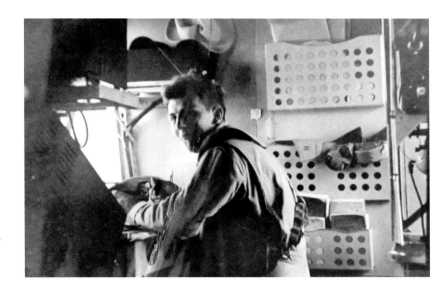

Lieutenant Commander Robert Bowser with maps.

Robert Bowser. The two navigators were Lieutenants Stanley Dutton and Charles Eadie, the Crew Chief was Chief Petty Officer Lee Steffen, who doubled-up on radar and radio. Meteorology and photography were handled by Petty Officer First Class William Dehn, the flight mechanics were Thomas Cox and James Burkett, the electrician was Carl Meyer, and Frank Maxymillion worked with Steffen as radioman and radarman. George Locklear, the rigger, doubled with Dehn as cook. The only civilian on board was Edgar Moore of Goodyear, who acted as a flight engineer alongside Lieutenant John Fitzpatrick. According to the US Navy they were a 'specially selected group of volunteers', but Frank Maxymillion recalls the selection process a little differently. He was taken before an executive officer, who said: 'OK Max, let's knock off the bullshit. Pack your ditty bag and climb aboard. And remember –we're all volunteers on this flight. Right?'

Accommodation for the men was certainly a step up from that in the old K-ships. The 83ft (25.3m)-long and 11.5ft (3.5m)-wide gondola was arranged with the pilots at the front, communications and radar stations just behind, and the fuel and ballast tanks, plus the twin engines, towards the back. The crew quarters were on the upper deck, which protruded within the envelope, and they had bunk beds, a galley for the preparation of hot food, a wardroom and even a small toilet.

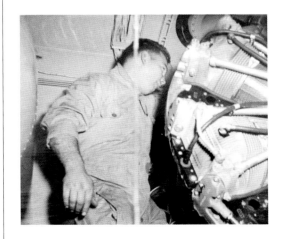

James Burkett services one of the engines.

Heading east and 600 miles (970km) out to sea, *Snow Bird* encountered a storm area with more snow falls and turbulent conditions, exactly the sort of weather for which the previous test flights had prepared them. By the second morning, 6 March, the airship was making good progress at first, with a southwesterly tailwind and the radar beams feeling out into the distance to avoid the mountainous peaks of the Azores. Flying at 1,000ft (300m), they were within the danger line. Although the weather reports were not that encouraging, *Snow Bird* was turned southward with engines at slow speed to save fuel. The following morning the wind conditions looked less favourable, and they changed course back to the east and initiated single-engine operation.

On the whole, life in the air soon settled into a routine, with the regular changes of watch and the day-to-day activities of the crew. It had been decided that they would not alter their clocks or watches during the flight, to make it easier to maintain watch schedules, meal times and so on. Even though the gondola was fairly noisy owing to the proximity of the big engines, the ride was gentle enough. It was much like being at sea aboard a small surface ship. To keep boredom at bay there were a number of diversions; spotting icebergs in the northern Atlantic or looking for whales and sharks. Addressing the Naval Airship Association on the fiftieth anniversary of the flight, radioman Frank Maxymillion recalled one encounter with whales:

I was standing a daylight watch when the pilot passed the word to look out the port side close in. I went forward to look out the windows on the flight deck and saw a pod of about seven or eight whales swimming along in the same direction we were going. We closed on them indicating that we were travelling very slowly or they were quick. They were always on the surface or just three or four feet below. We must have had them in sight for twenty to thirty minutes causing them no apparent concern. I guess they had nothing to fear from anything in the air.

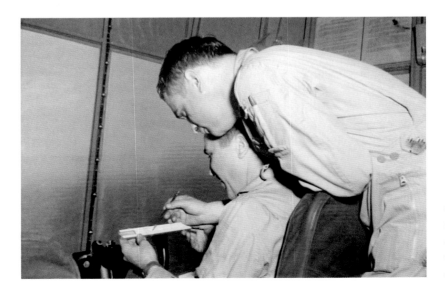

Lieutenants John Fitzpatrick and Stanley Dunton use a slide rule to make in-flight calculations.

Could it be that the whales had recognized in the airship a flying version of themselves? Best of all, the crew enjoyed coming into contact with other aircraft or vessels. During one of the night watches the copilot summoned Maxymillion to the flight deck to interpret a blinking light flashing at them from a mile or so ahead.

I grabbed the Aldus Lamp and sent him a Morse 'Q' signal, a radio operator's short-hand to repeat, which he began doing. I started mumbling letters, heck, I could just about copy Morse by hearing it let alone seeing it. I called topside to the bunkroom and summoned Chief Steffen. He came down the ladder in his skivvies and did an excellent job. It was a surface vessel asking 'What ship?' Just for the practice the chief handed me the lamp and told me to send: 'United States Airship *Snow Bird* on record-breaking cruise.'

As was to be expected, a few minor problems cropped up from time to time. The engines were spluttering or an air pressure alarm sounding, for example, but it was nothing this team could not handle. An engine would regularly be shut down and allowed to cool so that the flight mechanics could service it. Beyond the crew's influence, however, were the continuing headwinds. The plan had been for a twelve-day flight with a more moderate level of fuel consumption, but if the wind persisted there might not be enough fuel for the return leg without a refuelling stop, and Commander Hunt briefed the crew on the possibility of making a landing at Port Lyautey, Blimpron 14's wartime base. Accordingly the Super Constellation was directed to take the portable mast on ahead to make preparations for this contingency.

On the evening of 7 March they spotted

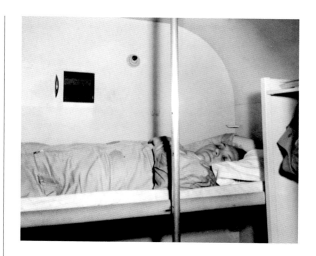

The upper deck of the ZPG-2 provided a relatively peaceful 'sackout' area between watches.

lights on the southwest tip of Portugal and the Algarve. *Snow Bird* then headed south towards the Canary Islands and on to Cape Verdi off the northwestern coast of Africa. They had successfully completed the first non-stop crossing by a non-rigid airship (the Blimpron 14 airships had made the flight in stages). Because the winds along the Portuguese coast were easing a little they were making better time, and by the morning *Snow Bird* had passed Casablanca on the coast of Morocco. As conditions looked better for the return leg, the decision was taken not to land at Port Lyautey after all. The entry in the ship's log reflected a renewed mood of optimism. 'Now we're committed...'

Flying parallel to the coast of Africa, the crew were able to benefit from far milder conditions than they had earlier experienced over the Atlantic:

It was so pleasant in fact that the pilots had the windows open on the flight deck. We were a few miles out but had a clear view of

A 1957 advertisement for Sperry flight control systems, featuring the long-duration flight of the Snow Bird.

Cape Yubi, the westernmost point of the Dark Continent. All I could make out of it was a gathering of huts and what appeared to be some low buildings, which may have been made of clay bricks. Everything appeared to be a dull, pale red. There was no evidence of any trees or other vegetation.

On its present heading the airship was approaching the Canary Islands from the northeast, and during the night watch they encountered fog banks which worried the pilot, as the islands' high peaks still lay in their path. He did not want to take the airship much higher, as this would entail venting helium, so he instructed Maxymillion to fire up the radar to navigate them through the islands. This equipment was only run when needed because it placed an additional burden on the engines.

We agreed that I would just give him directional orders, like 'easy left' or 'easy right'. He told me that if I waited long enough to have to give him a hard left or hard right order he would pitch me into the ocean without a life jacket and wait three days to pass the word about a man overboard... I never got a chance, or didn't want to leave the radar long enough, to go forward and look out the window to see how bad it was.

By the following evening, 9 March, they had passed Cape Verdi before heading west for the last 2,500-mile (4,023km) leg across the Atlantic all the way to Puerto Rico in the Caribbean.

As to be expected among a close-knit group of servicemen, there was the usual banter and joking around. Near the African coast Chief Steffen claimed he had captured a genuine African sea bat which had flown in through one of the open windows, and as each inquisitive junior crewman naively peeped into the empty cardboard box they received a hefty whack from a broom handle. 'I just figured Chief Steffen had too much time on his hands,' lamented Maxymillion. On 12 March William Dehn was presented with a surprise birthday cake, produced with some ingenuity, given the small galley's very limited cooking facilities. For some reason the Navy had not included birthday candles in the inventory, and instead they made do with some filter-tipped cigarettes tastefully poked into the cake.

By the ninth day of flight *Snow Bird* was closing in on some major aviation records, including the longest continuous flight, set at 200hr 12min by another US Navy airship three years earlier, and, more significantly, the *Graf Zeppelin*'s absolute distance record of 6,980 miles (11,233km), set in 1929 on the

Friedrichshafen to Tokyo leg of the round-the-world flight.

Snow Bird was getting progressively lighter as fuel was consumed, so it was decided to pick up sea water using a 50gal canvas bag lowered on a hoist at the back of the gondola. It was a tricky manoeuvre, flying at low level into the wind and throttling back the engines until the craft was virtually motionless. The crew could feel the airship settle at the stern as the full bag was hoisted up, and the pilot accelerated to regain some speed and height above the waves. The water was then pumped into an empty tank to act as ballast.

As *Snow Bird* neared the American coast there was a surge of excitement among the crew, who had been airborne for ten days. They had smashed the *Graf Zeppelin*'s record, and now they were heading home.

It had been determined that we would go in directly over Miami Beach. Being the middle of the day and coming in from the Atlantic we had to transition from the relatively cool air over the water to the hot, rising air over the sands of the beach. This caused the helium in the envelope to expand at an alarming rate, in turn causing the airship to start rising rapidly. The pilot's reaction to this was to nose the ship over to a very steep angle, apply max power to the engines, pump air into the ballonets and start, for the first time in the flight, to valve helium into the atmosphere.

From Miami Beach they turned west and out over the Everglades, flying low and slow to arrive in the vicinity of NAS Key West for a landing at sunset.

Just for the record

Snow Bird had been aloft for an unprecedented 264.2hr and had covered a track distance of 9,448 miles (15,200km). No other airship, or for that matter any other aircraft, had flown for so long and so far without refuelling.

At NAS Key West they were greeted by a large crowd of well-wishers. Pristine white navy hats were produced for the enlisted men before they could be paraded in front of Admiral William F. Halsey, who

Members of Snow Bird's crew being interviewed by the press, with Commander Jack Hunt in khakis.

congratulated them on behalf of President Eisenhower and presented Commander Hunt with the Distinguished Flying Cross. Hunt was later awarded the prestigious Harman Trophy by the President in person for his part in the flight of the *Snow Bird*. Young Frank Maxymillion and the regular crewmen were sent a medal and citation through the mail, and from Goodyear they each received a presentation model of the airship. 'All in all it was a great time. After a couple of days in the air I was glad I had been persuaded to "volunteer".'

It had been a spectacular flight by any measure, but the writing was on the wall for the US Navy's airships. An even bigger airship, the ZPG-3W, was wheeled out the following year in July 1958. It was the biggest non-rigid airship ever, and four were built, each one having a massive radar antenna within the helium envelope. But they were the last. With the advent of the satellite age the airship no longer seemed relevant, and despite the efforts of a few diehards within its ranks, the US Navy was forced to abandon its long tradition with lighter-than-air in 1962. Since then the Navy has continued to dabble in a number of evaluation studies using various different airship types, although these have been mainly short-lived episodes.

To date, no other airship has flown the Atlantic since the double crossing by *Snow Bird* over fifty years ago. Currently the record for the longest non-stop flight by any type of aircraft is held by Bertrand Piccard and Brian Jones of the *Breitling Orbiter 3*, which flew for 447hr and 55min in March

The forward cockpit section of Snow Bird's gondola is now displayed at the National Naval Aviation Museum in Pensacola, Florida, USA. (Ford Ross)

1999 when it made the world's first circumnavigation by balloon. Their rival balloonist and all-round record-collector, Steve Fossett, could not beat their endurance record, but he did establish a new distance record in 2006 when he flew the Burt Rutan designed *Virgin Atlantic Global Flyer* single-engined jet a distance of 25,766 miles (41,467km). This epic overlapping round-the-world flight included two Atlantic crossings, and although Fossett travelled further than any airship his greater speed meant the flight duration of 76hr 45min had not broken the airship records.

CHAPTER 16

Looking to the Future

～～

It is over fifty years since the US Navy's *Snow Bird* made its historic non-stop two-way flight. Looking back over the story of transatlantic airships, it is interesting to observe that the time between the first and last Atlantic crossing covers a period of just under forty years, while the golden era of the airships can be confined to a mere ten years or so, from the mid-1920s to the destruction of the LZ129 *Hindenburg* in 1937. This has not been for a lack of aspiration among the airship's faithful proponents, as countless schemes for ever-larger aerial leviathans have continued to fill the imagination and the pages of the popular science magazines. In a more recent phenomenon they have proliferated with the advent of the internet. The imminent return of the airship has been promised in every decade since the end of the Second World War, but this renaissance has seen an equal number of false dawns.

The British-built Skyships were arguably the world's first modern airships, constructed using the very latest lightweight materials and incorporating fly-by-wire controls and vectored thrust for improved manoeuvrability. (Airship Initiatives)

Could that be about to change, and will we see great transoceanic airships gracing the skies once more? The answer is a cautious 'possibly'.

A new generation of airships

Almost a decade into the twenty-first century and airships are still rare birds. When was the last time you saw one? Unless you live in the USA or are over eighty years old, the answer is probably very rarely. At present there are approximately thirty airships operating throughout the world, most of them in the USA, a figure that has hardly changed over the last thirty years. These are all relatively small craft with an average length of around 200ft (60m) or so, about a quarter that of the LZ129 or LZ130, and almost without exception they are non-rigid or pressure ships, their bloated profiles maintained solely by internal pressure. Goodyear kept the flag flying in the USA during the fallow years following the disbanding of the US Navy's airship division in the early 1960s, and its small fleet of blimps

bearing the distinctive winged-foot logo became iconic corporate ambassadors for the tyre company.

The Goodyear blimps do a great job. They have appeared over countless sports stadiums in the USA, and in 1972 the aptly named *Europa* was inflated in one of the giant hangars at Cardington in the UK before embarking on a number of tours of European countries. Undeniably, however, the Goodyear airships were old technology based on the pre-war L-ships, even incorporating their forebears' refurbished gondolas. Only the GZ-22 *Spirit of Akron*, completed in 1987, attempted to push the boundaries of airship technology for Goodyear, but its Allison 250-B17C turbo-props made it expensive to operate, and after an accident in 1999 the GZ22 was retired from service.

The first truly modern blimp had arrived on the scene in the late 1970s. The Skyship 500 was produced by the Cardington-based company Airship Industries, which had been founded as Aerospace Developments out of a project to design a colossal

The Zeppelin NT07 has a framework of three longitudinal girders, which means that the engines can be mounted on the sides of the hull, away from the passenger gondola. (John Albury)

A room with a view. The shores of Lake Constance seen from the rear panoramic window of the Zeppelin NT07.

97,000,000cu ft (2,750,000 cu m) Gas Carrier for Shell International Gas Limited. When the Gas Carrier failed to materialize, the design team, headed by former naval architect Roger Monk, developed the Skyship series. The 500 and the larger 600 feature many of the elements now considered standard for small airships, including lightweight materials for the envelope, gondola and control surfaces, fly-by-wire controls and, perhaps most importantly, a system of tilting or vectoring the propellers to direct thrust for improved manoeuvrability. Their main role was primarily seen as aerial billboards, although they were also promoted for airborne surveillance and maritime patrol duties, as well as for tourism. Unfortunately, while Airship Industries may have created a superb airship, it failed to find sufficient customers, and the company folded in 1989. The rights to manufacture and operate the Skyship series now lie with Airship Management Services in the USA, and a number of their airships continue to fly in advertising and security roles in various parts of the world.

The latest version of the Skyship 600B has a volume of 247,500cu ft (6,666cu m) and is powered by either Porsche 930 turbocharged piston engines or, in some cases, the Textron Lycoming IO-540. Their eye-in-the-sky capabilities have been put to good use at several Olympic Games, including two Skyships deployed over Athens in 2004. Other non-rigid designs have also come along, most notably the highly successful Lightships, built by the American Blimp Corporation and distinguished by a translucent envelope which can be lit up internally at night like a giant flying light bulb.

Return of the Zeppelins

Meanwhile, in Germany, a slumbering giant of the airship world was preparing to re-enter the fray in the 1990s. Following the end of the war the Zeppelin company had survived by diversifying into other forms of engineering. Now, as Zeppelin Luftschiff Technik (ZLT), it was proposing to build airships once more, and not just blimps, but rigid framed airships. The Zeppelin NT07 (NT for *Neu Technologie*) was created by updating the original Zeppelin concept using modern materials to create a framework comprising three aluminium

Passengers aboard the Zeppelin NT enjoy superb views through the gondola's large windows. By 2008 NTs were operating 'flight-seeing' trips in Germany, Japan and the USA. (Zeppelin Tours)

longitudinal girders running the length of the airship, connected transversely by a series of graphite-reinforced triangular cross-frames. This provides attachment points for two Textron Lycoming IO-360 engines mounted directly to the hull sides, with a third engine on the tail for vastly improved ground manoeuvrability. The NT07's 290,000cu ft (8,255cu m) volume makes it the largest manned airship for the present. The prototype first flew in 1997, and since then the company has built three more, now operating at Friedrichshafen in Germany, the Tokyo area of Japan and over San Francisco Bay in the USA.

One disadvantage of the rigid airship is the difficulty of delivering it to a distant location such as Japan or the USA. The NT07's 246ft (75m) rigid frame makes it impossible to pack it for shipping and rein-flation at the intended destination, as would be done with a non-rigid. When ZLT sold an NT07 to Japan in 2004 the original plan had been to fly it all the way there via Finland and across Russia, roughly following the round-the-world route of the *Graf Zeppelin*. But in the event the appropriate permissions for overflight were rescinded, and instead it was carried on a special cargo ship designed for the transportation of large objects such as big yachts. Getting the fourth NT07 to San Francisco in 2008 raised the real hopes of a transatlantic flight via the Azores, as Brian Hall of the operating company, Airship Ventures, explained:

> We had looked at 'kiting' – towing behind a fast ship – with underway replenishment and extended range fuel tanks. In the end, it was the schedule that dictated our choice. The favourable weather window when the wind drops to near zero around the Azores was June, and as the ship had only made its first flight in May there was no way we could have hit that weather window to fly it to the USA. It became a choice of transfer by boat in 2008, or fly it in 2009.

They opted for delivery by surface vessel, and the Airship Ventures Zeppelin arrived at Beaumont, Texas, on 14 October 2008. It was then flown to its new base at Moffett Field near San Francisco, former home of the US Navy's *Macon*. This was the first Zeppelin flight in the USA in seventy-one years. On 21 November 2008 it was christened *Eureka*, which had been the project name for the transatlantic delivery, and it is currently providing tourist 'flightseeing' trips over the San Francisco area.

Looking ahead, ZLT is considering an extended version of the NT07, and plans exist for a nineteen-seat NT14. A European group called Zeppelin Europe Tourismus or 'Tourism' (ZET) is also promoting the idea of an even bigger Zeppelin to be known as the 'Z' ship, capable of carrying forty passengers between major European cities and on pleasure flights or cruises. It is likely to be several years before this becomes available, so what are the prospects for a return of transatlantic passenger flights? Frankly, not very good in the foreseeable future, as this would require a far bigger airship with an even bigger development budget. There is also the question of whether there is a market for an expensive and slow luxury crossing in these fast-moving times, when a jet aircraft can do it in a matter of hours rather than days, not to mention the costs of developing and certificating a new airship. It is far more probable that we will continue to see incremental steps with a range of improved airship types, followed only later by the larger passenger carriers.

This Zeppelin NT07 was transported by sea to Japan in 2005 aboard the special heavy transport ship Dock Express 10.

Engineers work on the tail framework of the fourth Zeppelin NT07 at the company's shed in Friedrichshafen. Transported across the Atlantic in 2008, this airship is now operating in the USA.

ABOVE: The Eureka!, the latest Zeppelin NT, is now in the USA, providing tourist flights over the San Francisco Bay area. (Roger Cain – Airship Ventures)

Heavy-lift and hybrids

Aside from passenger operations, the potential of airships to transport freight or large indivisible loads is attracting considerable interest, and, following on from the proposals for vast airships advanced in the 1950s and 1960s, the notion that bigger is better has continued to inspire. In the mid-1970s the Airfloat project, headed by Edwin Mowforth of the University of Surrey, outlined a range of freight-carrying airships, the largest of which, the Airfloat HL, was capable of transporting up to 400 tons beneath its 1,345ft (400m) hull. Power for this colossal airship was to be provided by six Proteus gas turbines driving 21ft (6.4m)-diameter propellers. As the airship was too big to fit within any existing shed it was proposed to construct it in sections in the open, using a turntable system. In addition to moving freight, Mowforth's team also devised the Aerotel version, able to transport a hotel module with accommodation for around 250 passengers, possibly across the Atlantic.

Neither version of the Airfloat airships ever went into production, but they might have provided the inspiration for the CargoLifter project established in 1996, with Mowforth as a member of the design team. Some funding came in the form of regeneration grants from the German government and from the issue of shares. The plan was to build the CL160, a semi-rigid that was going to be bigger than the *Hindenburg* and capable of transporting loads of up to 160 tons to any location. Amid much hoo-ha a vast shed, shaped like a massive jelly mould, was erected at a former East German airfield at Briesen-Brand, just to the south of Berlin. In addition, a visitors' centre known as CargoLifter World was opened, and glossy brochures and newsletters spread the word. CargoLifter built a small test airship and a massive spherical balloon, and even purchased a Skyship for pilot training, but the one thing it did not do was build the big CL160 airship. When the bubble burst in 2002 the company went into liquidation and

ABOVE: A Z-ship, the vision of the Zeppelin Europe Tourismus organization, which has plans for a network of bigger airships connection major European cities. (ZET)

RIGHT: The spacious interior of the Z-ship's gondola. When it comes to designing airships, accommodation space is plentiful while weight is the more important factor. (ZET)

LEFT: The rear end of the Z-ship's gondola incorporates a wide window, just as with the NT07, plus a glass floor for the braver tourists. (ZET)

CargoLifter World
Experience Zero Gravity.

www.cargolifter-world.de

Registriertes Projekt
der Weltausstellung EXPO2000 HANNOVER

A ticket to CargoLifter World was not a ticket to ride. The massive shed was completed at Briesen Brand, on the site of a former East German airfield to the south of Berlin, but the big CL-160 heavy-lift airship was never built.

The scale of the CargoLifter shed is almost beyond comprehension, but if you look carefully there is the tiny figure of a man standing beside the door opening.

the jelly-mould hangar was turned into an indoor tropical holiday resort.

Despite this high-profile failure, other schemes for freight hauling by airship may yet get off the ground. In northern Canada, for example, rising temperatures are causing the network of ice roads to soften. These provide a vital link to the many remote communities and the mineral mines, but building conventional roads is impossible, which is why the airship is seen by many as a solution. Some experts are predicting a demand for over 100 hybrid airships for Canada, each one capable of transporting 50-ton payloads.

In parallel with the traditional blimps, a number of unconventional designs have also emerged. These are referred to as hybrids, because they combine the buoyancy of helium lifting gas with either helicopter-style rotors or with the aerodynamic lift generated by a wing- or aerofoil-shaped hull. The latter are known as 'lifting bodies', and some concepts resembling flying saucers while others are

This proposal for a modern 590ft (180m) rigid framed airship was put forward by Rigid Airship Design of the Netherlands in the late 1990s. (Rigid Airship Design)

flattened versions of the blimp. Several companies have put forward concepts for a lifting-body airship, most notably the post-Airship Industries team, which has undergone a number of changes over the years and now trades as Hybrid Air Vehicles (HAV). Other companies working on similar projects include Worldwide Aeros, RosAero Systems in Russia, Lockheed Martin and Millennium Airships in the USA, plus countless others to be found on the internet.

Many of these schemes were given impetus, credibility, and in some cases hard cash in the form of development funds as a result of the US Defense Advanced Research Projects Agency's (DARPA) heavy-lift air vehicle programme, known as 'Walrus', launched in 2005. This called for a massive

The SkyCat is a lifting-body concept designed principally for the transportation of large loads, although it could be produced as a people carrier too. (Hybrid Air Vehicles)

Routine maintenance is carried out on a Zeppelin at Friedrichshafen once again, seventy years after the great transatlantic Zeppelins grace the skies.

So far only Lockheed Martin has produced a flying manned prototype of a lifting-body airship, the P-791, which made its maiden flight in 2006. HAV says its SkyCat 50, designed to carry up to 50 tons of freight or a corresponding number of passengers, could be flying by 2011 or 2012, and will be followed by the bigger 100- or 200-ton version. Worldwide Aeros is proposing to develop a similar lifting-body airship, with a 210ft (64m)-long luxury 'sky yacht' version, the ML866, complete with lounge areas, observation deck and state rooms. For most of these projects transatlantic services are not necessarily a priority, but could become a possible supplementary application.

A new age of the airship?

While the airship will never be the solution to all our transport needs, it is being viewed very seriously as the ideal vehicle to perform a number of important tasks, and for the first time in decades it seems that lighter-than-air transport may have a serious future. In many ways, as we stand at the dawn of the second century of flight, the story of the transatlantic airships has come full circle. The visionaries painting their pictures of incredible flying machines continue to provide us with a glimpse of a future that might yet be, one in which the shadows of giant airships fall once more upon the dark Atlantic waves.

Keep watching the skies.

hybrid airship capable of moving 500 tons direct from 'fort to fight' to transport a rapid-response force to any given trouble spot in the world without the delays inherent in conventional surface vessels. Although funding for Walrus was terminated in 2007, several proposals for large passenger carrying airships have emerged as offshoots from the initial studies for the military.

Chronology

1900 **2 July:** First flight of Count Zeppelin's LZ1 rigid framed airship.

1901 **19 October:** Alberto Santos-Dumont successfully flies his little No.6 airship around the Eiffel Tower to claim the Deutsche Prize.

1909 **25 July:** Louis Blériot crosses the English Channel in his monoplane.

16 November: The DELAG transport company is established to conduct passenger flights between German cities; they commence the following year.

1910 **15 October:** Walter Wellman's semi-rigid *America* departs from Atlantic City. It flies for 64hr and covers over 900 miles before ditching in the Atlantic. Its six-man crew and cat are picked up by RMS *Trent*.

16 October: The *Clemant-Bayard II* makes the first crossing of the English Channel by airship, flying from France to England.

4 November: Ernest T. Willows makes the first England-to-France Channel crossing by airship in the Willows III *City of Cardiff*.

1910–12 The *Suchard* non-rigid is com-pleted in 1910 for Josef Bruckner's intended flight from Tenerife to South America. The airship is length-ened after trials, but the funding runs out before the attempt is made.

1912 **2 July:** Shortly after taking off from Atlantic City, Vaniman's *Akron* airship falls into the sea, killing all on board.

1917 **21–15 November:** The L59 (LZ104) sets a new distance record of 4,200 miles (6,800km) when it flies with the intention of reinforcing troops in German East Africa, but turns back before reaching its destination.

1919 **15 May:** The US Navy's plans to make an Atlantic crossing via the Azores with the C-5 non-rigid are scuppered when sudden gales rip the airship from the handling party at St Johns, Newfoundland, and it disappears over the Atlantic.

14–15 June: Alcock and Brown make first non-stop transatlantic flight, from Newfoundland to Ireland, in their Vickers Vimy aeroplane.

2–6 July: Less than a month after Alcock and Brown's flight, the crew of the British airship R34, under Commander Scott, make the double Atlantic crossing, from Edinburgh to Mineola, Long Island, flying back to the UK on 9-13 July and landing at Pulham, Norfolk.

1921 **24 August:** The British-built R38 breaks up over the Humber during acceptance flights before it can be handed over to the US Navy as the ZR-2. Forty-four are killed, five survive.

1924 **12–15 October:** Hugo Eckener takes the Zeppelin LZ126 on a one-way delivery flight from Friedrichshafen to Lakehurst, New Jersey. The US Navy names it ZR3 *Los Angeles*.

1927 **20 May:** Charles Lindbergh becomes the first person to fly the Atlantic solo non-stop in his Ryan NYP monoplane *Spirit of St Louis*, a flight of 33.5hr from New York to Paris. According to some sources this makes him the 92nd person to fly across the Atlantic.

1928 **18 September:** First flight of the LZ127 Graf Zeppelin.

11 October: First transatlantic flight of the *Graf Zeppelin*, flying from Friedrichshafen to Lakehurst.

1929 **8–29 August:** The *Graf Zeppelin* makes an epic round-the-world flight, covering 20,500 miles.

1930 **29 July–1 August:** The R100 flies from Cardington to Montreal, returning to the UK on 13–16 August.

5 October: The R101 crashes at Beauvais in France; six survivors and forty-eight dead. Consequently the R100 is broken up the following year and sold as scrap.

1936 **4 March:** The official maiden flight of the as yet unnamed LZ129.

31 March: The LZ129 *Hindenburg*

sets out on its first Atlantic crossing, on a round trip to Rio de Janeiro in South America.

1937 **6 May:** The *Hindenburg* is engulfed in flames as it comes in to moor at Lakehurst after the first crossing of the 1937 season. The *Graf Zeppelin* is in the air, returning to Germany from South America at the time of the accident; the last transatlantic flight of the giant Zeppelins.

1938 **14 September:** The LZ130 *Graf Zeppelin II*, sister ship to the *Hindenburg*, makes its first flight. It never carries paying passengers, and after only thirty flights is broken up, along with the first *Graf Zeppelin*, in early 1940.

1944 **30 May–1 June:** The US Navy's K-130 and K-123 make the first transatlantic flight by non-rigid airships, departing from Argentia in Newfoundland and flying via the Azores to French Morocco. They are the first of eight K-ships of the ZP-14 blimp squadron, which serves in North Africa and Europe in the latter stages of the Second World War.

1945 **28 April–1 May:** The final two US Navy ZP-14 blimps, K-89 and K-114, cross the Atlantic, flying to French Morocco. Following the end of hostilities they serve in mine-spotting duties until ZP-14 is decommissioned on 31 January 1946.

1957 **4–15 March:** The US Navy's ZPG-2 *Snow Bird* makes a record-breaking 264.2hr non-stop unrefuelled flight when it crosses the Atlantic to the coast of Africa and then flies back to the USA, covering a total distance of 9,448 miles. This is the last transatlantic airship flight to date.

Bibliography

Abbott, Patrick, *Airship – The Story of the R34*, Adams & Dart Jupiter Books, 1973.

Allen, Hugh, *The Story of the Airship*, Goodyear Tire & Rubber Co, 1931.

Allen, Peter, *The 91 Before Lindbergh*, Airlife, 1984.

Althoff, William F., *USS Los Angeles – The Navy's Venerable Airship and Aviation Technology*, Brasseys, 2004.

Archbold, Rick, *Hindenburg – An Illustrated History*, Warner Books, 1994.

Asquith, Tom, and Deacon, Kenneth, *Howden Airship Station 1915–1930*, Howden Civic Society 2006.

Bauer, Manfred, and Duggan, John, *LZ130 Graf Zeppelin and the end of Passenger Airships*, Zeppelin Museum Friedrichshafen, 1994

Bentele, Eugen, *The Story of a Zeppelin Mechanic*, Friedrichshafen, 1992.

Botting, Douglas, *Dr Eckener's Dream Machine*, Harper Collins, 2001.

Brewer, Griffith, and Alexander, Patrick, *Aeronautics: An Abridgement of Aeronautical Specifications Filed at the Patent Office from AD 1815 to AD 1891*, Taylor and Francis, 1893 (reprinted by Boekhandel en Antiquariaat, Amsterdam, 1965).

Brooks, Peter W., *Zeppelins: Rigid Airships 1893–1940*, Putnam 1992.

Burney, Dennistoun, *The World, The Air & The Future*, Alfred A. Knopf, 1929

Chamberlain, Geoffrey, *Airships – Cardington*, Terence Dalton, 1984.

Countryman, Barry, *R100 in Canada*, Boston Mills Press, 1982.

Christopher, John, *Riding the Jetstream – From Montgolfier to Breitling*, John Murray, 2001.

Dick, Harold G., with Robinson, Douglas H., *The Golden Age of the Great Passenger Airships Graf Zeppelin & Hindenburg*, Smithsonian Institution Press, 1985.

Duggan, John, and Cord Meyer, Henry, *Airships in International Affairs, 1890–1940*, Palgrave, 2001.

Eckener, Hugo, *My Zeppelins*, Putnam, 1958.

Griehl, Manfred, and Dressel, Joachim, *Zeppelin! The German Airship Story*, Arms & Armour, 1990.

Hartcup, Guy, *The Achievement of the Airship*, David & Charles 1974.

Hedin, Robert, *The Zeppelin Reader*, University of Iowa Press, 1998.

Jamison, T.W., *Icarus Over the Humber – The Last Flight of Airship R38/ZR-2*, Lampada Press, 1994.

Johnston, E.A., *Airship Navigator – One Man's Part in the British Airship Tragedy 1916–1930*, Skyline, 1994.

Kirschner, Edwin J., *The Zeppelin in the Atomic Age*, University of Illinois, 1957.

Knausel, Hans G., *Zeppelin and the United States of America*, Luftschiffbau Zeppelin, 1981.

Lehmann, Captain Ernst, and Mingos, Howard, *The Zeppelins*, Putnams, c1927.

Litchfield, P.W. and Allen, Hugh, *Why Has America No Rigid Airships?*, 7 C's Press, 1976.

Mabley, Edward, *The Motor Balloon 'America'*, Stephen Greene Press, 1969.

Maitland, Air Commodore E.M., *The Log of HMA R34 – Journey to America & Back*, Hodder & Stoughton, 1920.

Meager, Captain George, *My Airship Flights 1915–1930*, William Kimber, 1970.

Meyer, Henry Cord, *Airshipmen Businessmen and Politics 1890–1940*, Smithsonian Institution Press, 1991.

Nielsen, Thor, *The Zeppelin Story*, Wingate, 1955.

Pratt, H.B., *Commercial Airships*, Nelson, 1920.

Robinson, Douglas H., *Giants in the Sky – A History of the Rigid Airship*, Foulis, 1973.

Robinson, Douglas H., *LZ129 Hindenburg*, Morgan Aviation Books, 1964.

Robinson, Douglas H., *The Zeppelin in Combat*, G.T. Foulis, 1962 (reprinted by Schiffer Military History, 1994).

Rosendahl, Commander C.E., *What About the Airship?*, Charles Scribner's Sons, 1938.

Shock, James R., and Smith, David R., *The Goodyear Airships*, Airship International Press, 2002.

Shock, James R., *US Navy Pressure Airships 1915–1962*, Atlantis 1993.

Shute, Nevil, *Slide Rule – The Autobiography of an Engineer*, Heinemann, 1954.

Topping, Dale, (edited by Brothers, Eric), *When Giants Roamed the Sky*, University of Akron, 2001.

Vaeth, J. Gordon, *Graf Zeppelin*, Muller, 1959.

Vaeth, J. Gordon, *Blimps & U-Boats – US Navy Airships in the Battle of the Atlantic*, United States Naval Institute, 1992.

Vissering, Harry, *Zeppelin – The Story of a Great Achievement*, 1922.

Walmsley, Nick le Neve, *R101 – A Pictorial History*, Sutton Publishing, 2000.

A number of journals, magazines and newspapers were also consulted, including *Airship*, *Airshipworld*, *Buoyant Flight*, *Die Woche*, *Dirigible*, *Icare*, *National Geographic*, *The Noon Balloon*, *Popular Mechanics* and *Popular Science*.

Index